Philosophical Perspectives on Safety and Risk

From Harm to Harmony

Mohammad Yazdi

CRC Press
Taylor & Francis Group
Boca Raton London New York

CRC Press is an imprint of the
Taylor & Francis Group, an **informa** business

Designed cover image: Shutterstock

First edition published 2027
by CRC Press
2385 NW Executive Center Drive, Suite 320, Boca Raton FL 33431

and by CRC Press
4 Park Square, Milton Park, Abingdon, Oxon, OX14 4RN

CRC Press is an imprint of Taylor & Francis Group, LLC

ISBN: 9781041062882 (hbk)
ISBN: 9781041051947 (pbk)
ISBN: 9781003634720 (ebk)

DOI: 10.1201/9781003634720

Typeset in Times
by Deanta Global Publishing Services, Chennai, India

Contents

About the Author

Mohammad Yazdi's academic and professional journey began with a Bachelor of Science in Safety and Technical Protection Engineering from the Petroleum University of Technology in Abadan, Iran, awarded in 2012. He later obtained a Master of Science in Industrial Engineering from Eastern Mediterranean University in Famagusta, Cyprus, in 2017.

Following his master's studies, Mohammad joined the Centre for Marine Technology and Ocean Engineering (CENTEC) at the University of Lisbon, Portugal. He contributed to research activities on-site during 2017 and 2018 and continued his collaboration remotely in 2019.

In 2022, he commenced a dual PhD program through a collaboration between the Centre for Risk, Integrity, and Safety Engineering (C-RISE) at Memorial University of Newfoundland, Canada, and Macquarie University in Australia.

Prior to his academic career, Mohammad worked in industry between 2012 and 2016 in roles including Firefighter, Safety Officer, WHS/OHS Advisor, and Auditor. His experience covers sectors such as power generation and the oil and gas industry.

His research and professional work focus on system safety, risk assessment, resilience engineering, process integrity, and asset management, particularly in renewable and non-renewable energy infrastructure. He has experience contributing to large-scale and technology-intensive projects, supporting asset operators, developers, and maintainers.

His work involves linking academic research with industry applications by developing approaches that support decision-making, systems thinking, and problem-solving. He applies an asset lifecycle perspective that spans capital investment planning, operation, maintenance, and decommissioning, with attention to functionality, reliability, and business outcomes.

Preface

Safety is often described as a technical discipline, a regulatory requirement, or a managerial function. Yet beneath procedures, standards, and metrics lies a deeper question: *How do we know what is safe, and how should we decide in conditions of uncertainty?* This book begins from the premise that safety is not merely an engineering outcome but a philosophical, ethical, and epistemological undertaking. It is shaped by how we understand knowledge, responsibility, technology, power, culture, and the future.

Across industries and societies, risk is no longer confined within clear boundaries. Digital platforms reshape labor and finance; emerging technologies blur the line between cyber and physical harm; organizational cultures quietly normalize deviance; global safety norms encounter diverse cultural realities; and energy systems evolve into tightly coupled socio-technical-ecological networks. In this environment, traditional harm-avoidance models, valuable though they remain, are no longer sufficient. What is required is integration: integration of knowledge sources, integration of ethical principles, integration of governance mechanisms, and integration of system perspectives.

This volume brings together philosophy and practice to advance such integration.

Chapter 1 lays the epistemological foundation, tracing how classical and modern theories of knowledge inform real-world safety decision-making. By bridging Plato, Aristotle, and Kant with lived professional experience, it demonstrates that rigorous safety practice depends not only on data, but on reflective judgment and practical wisdom.

Chapter 2 expands the lens to risk governance in the digital age. It identifies recurring mechanisms of harm, information asymmetry, misaligned incentives, behavioral vulnerabilities, weak accountability, and distills universal principles applicable across finance, digital platforms, and safety systems. The aim is a domain-agnostic framework for governing risk in complex socio-technical environments.

Chapters 3 and 5 examine the ethical integration of emerging technologies. As AI, automation, robotics, digital twins, and wearable systems reshape safety practice, new forms of risk accompany new capabilities. These chapters introduce the concept of technological ethos, arguing that innovation must be guided by dignity, transparency, accountability, and human oversight if it is to strengthen rather than undermine safety.

Chapter 4 turns inward to organizational life, applying the concept of the "banality of evil" to workplace health and safety. It explores how ordinary routines, hierarchical pressures, and normalized shortcuts can accumulate into systemic harm, and how ethical leadership and accountability are central to reversing such trajectories.

Chapter 6 addresses the tension between universal safety norms and cultural particularity. It argues that while certain ethical commitments may be global, effective safety strategies must resonate with local values, risk perceptions, and institutional contexts. Philosophical reflection becomes a bridge between global narratives and lived realities.

Finally, Chapter 7 synthesizes these strands into a comprehensive framework for integrated safety in the energy sector. Moving from harm to harmony, it conceptualizes safety as an emergent property of complex systems, grounded in ethical principles and strengthened through epistemological pluralism. It offers practical pathways, integrative risk assessment, participatory simulation, adaptive governance, resilience dashboards, and ethical impact assessment, for operationalizing this vision.

Together, these chapters present safety not as a static endpoint, but as an evolving societal endeavor. They invite engineers, policymakers, scholars, and practitioners to reconsider safety as a shared philosophical commitment, one that balances evidence with wisdom, innovation with responsibility, and efficiency with justice.

The journey from harm to harmony is neither simple nor complete. It is a continuous process of questioning assumptions, learning across boundaries, and aligning technological progress with human values. This book offers both a conceptual compass and a practical roadmap for that journey.

M. Yazdi
Sydney, March 2026

1 Epistemological Approaches to Safety Knowledge and Decision-Making

Bridging Theory and Practice

INTRODUCTION

Safety in engineering should not be reduced to following procedures or maximizing practicability; it is fundamentally about understanding what is known and what remains uncertain. Understanding safety requires asking: How can something be determined as safe? How should beliefs about risk guide decisions? These are epistemological questions, questions about the nature and sources of knowledge, that echo debates from ancient philosophy. This chapter bridges those centuries-old ideas with the realities of modern industrial safety, demonstrating how philosophical concepts such as truth, belief, evidence, and doubt directly shape how engineers interpret data, assess risks, and learn from incidents.

In weaving together theory and practice, personal reflections illustrate how experience reinforces epistemological lessons. Flashback stories, from a childhood experiment gone awry to a critical hazard identified on a job site, show how lived experience applies these principles. These narratives make the abstract concrete: a childhood engineering curiosity demonstrates that belief without proof can be dangerous, echoing Plato's notion that knowledge requires both truth and justified belief. Later, in professional practice, it becomes clear that following rules alone is insufficient; practical wisdom, as Aristotle described in his concept of phronesis, is equally essential.

Our goal is to provide a comprehensive overview of how theories of knowledge (from Plato and Descartes to Dewey and contemporary thinkers) inform safety knowledge and decision-making [1]. Each philosophical insight will be connected to the world of engineering practice, using stories to illustrate how theoretical ideas play out on the shop floor or in the control room. In doing so, we argue that cultivating a philosophical mindset is crucial: it teaches us to balance evidence and intuition, to value reflection as much as data, and ultimately to make safer decisions.

DOI: 10.1201/9781003634720-1

PHILOSOPHICAL FOUNDATIONS OF KNOWLEDGE

Epistemology, the study of knowledge, addresses fundamental questions: "What is knowledge?" [2] and "How can claims to knowledge be justified?" Ancient philosophers established the foundation for these inquiries. Plato defined knowledge as the intersection of truth and belief: to truly know something, one must both believe it and have it be true, supported by justification. This classic "justified true belief" model has direct relevance for safety: merely believing that a practice is safe is insufficient unless the belief is supported by evidence and truth alone does not constitute knowledge without justified belief.

Aristotle expanded on this by distinguishing types of knowledge [3], emphasizing that ethics and safety are rooted not merely in universal rules but in practical wisdom, or phronesis. For Aristotle, knowing the right course of action, such as performing safely, cannot be learned from books alone; it requires experience and character. He observes, "Practical wisdom cannot be acquired solely by learning general rules. We must also acquire, through practice, those skills that enable us to put our general understanding into practice in ways that are suitable to each occasion." In modern terms, this implies that a safe worker requires more than procedures; judgment is needed to apply rules effectively in context. Safety decisions often involve ambiguous situations, and Aristotle's insight underscores that genuine wisdom arises from repeated practice and experience, not merely from memorizing abstract formulas.

We inherit a rich legacy of later epistemology as well. Descartes (seventeenth century) taught us the value of doubt in seeking certainty [4]. He famously set out to doubt everything that could be doubted, to see what remains indubitable. In safety, a Cartesian spirit urges skepticism of our assumptions: for example, not simply trusting a system is safe because "we've always done it this way"; however, questioning whether unnoticed failures or biases lurk [5].

Kant (eighteenth century) later synthesized rationalism and empiricism [6]. He argued that the human mind imposes structure on raw sensory data. In practical terms, Kant reminds us that safety knowledge arises both from experience (observing incidents, tests) and from the conceptual framework our mind uses to interpret data. In his own words, he "synthesized early modern rationalism and empiricism." This balance is evident when an engineer uses mathematical models (rationality) informed by sensor data (empiricism) to predict risk. Kant also emphasized that reason gives us principles (like moral imperatives) while experience supplies the content, analogous to how safety principles guide us; however, we need real-world data to apply them responsibly.

Even more recent thinkers emphasized active learning. For example, John Dewey defined knowledge as the "conscious and voluntary effort to establish belief upon a firm basis of reasons" [7]. Dewey's definition captures the spirit of scientific inquiry: to *make* a justified belief by reasoning from evidence, not just to accept beliefs passively. In safety practice, this translates to a mindset of continuous questioning: gathering data, analyzing incidents, and constantly updating our "belief" in whether something is safe.

A childhood flashback illustrates these ideas. While building a small go-kart in the family garage, the wooden frame appeared sturdy, and it was believed to be safe to race down a hill. When the axle snapped under load, it became clear that belief without proof can fail catastrophically. This early experience exemplifies Plato's insight: the frame was not truly safe until the belief was verified. In that moment, the principle of "justified true belief" was lived; knowing something is safe requires both belief and confirmation. The incident left a lasting impression, highlighting the importance of testing and validating assumptions in every project.

From a professional standpoint, the deeper lesson is that safety relies on justified true belief. An engineer may *believe* a bridge design is robust, but only when testing (or calculation) shows *truth* can we claim knowledge. Aristotle's warning also resonated: had we only followed some "general rule" (e.g., "use wood for frames"), without the practical wisdom of testing under realistic loads, our rules failed us. It was a visceral introduction to why theory alone cannot guarantee safety; practice and evidence must accompany every design.

FROM THEORY TO SAFETY PRACTICE

Having reviewed how philosophers conceptualized knowledge, we now consider how these ideas manifest in the field of safety science. Safety management itself can be seen as a type of knowledge-generation process. As risk scholar Aven notes [8], "safety science, implies a safety knowledge generating process, comprising two components, (i) knowledge about safety-related phenomena, and (ii) conceptual tools (concepts, theories, principles and methods) to understand and manage safety." In other words, we gather *facts* about incidents and hazards *and* develop theories and methods to interpret and act on those facts. This dual aspect mirrors Kant's insight: raw data (empirical knowledge of hazards) must be structured by our conceptual framework (models of risk, safety principles).

Modern epistemology emphasizes that knowledge is never absolute. Every safety judgment carries epistemic uncertainty, uncertainty about what is truly known. Niklas Möller argues that a robust concept of safety must explicitly account for this uncertainty [9], defined as "the degree to which we are uncertain of our knowledge of the situation at hand." In practice, this requires asking questions such as: How confident are we that this procedure will work? What aspects of this system remain unknown? A common pitfall is overconfidence in data, for example, assuming that "since no crashes have occurred, the plant must be safe." Möller emphasizes that safety involves not only reducing risk, in terms of probabilities and severities, but also actively addressing uncertainty by recognizing potential blind spots.

The notion of epistemic doubt is illustrated by real incidents. At a chemical plant, a junior safety engineer observed a situation where an experienced operator claimed that a particular valve was "inherently safe," having always failed closed during tests. The operator's belief assumed immovability under pressure, yet theoretical training and careful skepticism prompted a more thorough inspection, including an unscheduled stress test. The valve failed open in one trial, revealing that prior assumptions did not reflect reality. Knowledge based on limited observations proved incomplete; broader testing expanded the "knowledge base" and exposed a hidden hazard. This episode reinforces Aristotle's lesson that reliance on routine or hearsay is insufficient; practical wisdom requires verifying what others take for granted. It also aligns with Rushby's assertion that a central challenge in safety cases is reducing epistemic doubt [10]. Testing assumptions enabled more reliable knowledge and closed the gap between belief and truth.

Safety professionals often speak of different types of knowledge. It is helpful to distinguish at least three categories:

- **Explicit Knowledge:** Formalized information like technical manuals, standards, accident reports, and training curricula. This is propositional knowledge we can write down and teach (e.g., "The tensile strength of this alloy is 250 MPa.").
- **Tacit Knowledge:** Personal know-how and intuition that are hard to codify. For example, an experienced welder might "feel" a joint is cool without a temperature gauge. Polanyi famously noted that "we can know more than we can tell," pointing to the tacit dimension of expertise. In safety, tacit knowledge includes instinctive hazard recognition or an engineer's sense for a faulty design that defies written explanation.
- **Collective Knowledge:** The shared culture and values of an organization. This includes unwritten norms (e.g., "double-check everything even if you trust your colleague") and lessons learned from past incidents that become part of the corporate memory. Collective knowledge is how societies and teams embed safety practices.

These components interact: explicit rules guide action, yet they are effective only when applied with tacit understanding and reinforced by a positive safety culture. For example, a procedure may instruct workers to wear harnesses; however, compliance is ensured only when an experienced foreman emphasizes the rule and a culture discourages shortcuts. The interplay between explicit, tacit, and collective knowledge is critical: formal procedures provide structure, tacit knowledge enables judgment in context, and cultural norms sustain consistent practice. Consider a university engineering project designing a small-scale fluid system. Students had access to extensive formulas and theoretical models, yet the professor continually challenged them: "That works in your textbook case; consider what happens if the pipe is not perfectly level or if the fluid is more viscous." This approach trained students to anticipate real-world variability, cultivating phronesis, practical wisdom to apply theory flexibly under real conditions. Dewey's principle of actively validating beliefs was central: knowledge must be tested against lived experience, not assumed from idealized models. This combination of explicit rules, contextual judgment, and cultural reinforcement remains essential in industrial safety, ensuring that procedures are not merely followed; rather, they are applied wisely and effectively.

In complex situations (an accident investigation, for example), we inevitably confront gaps in knowledge. Formal reasoning can only go so far. Rushby observed that in a safety case, the reasoning about design can be automated via logic; however, *"the first [concern] is a form of epistemology and requires human experience and insight."* In other words, deciding *what we know and what we are uncertain about* is a human judgment call. Automation and checklists help; however, ultimately humans must interpret signs and learn from anomalies. When our plant experienced a near-miss due to a sensor malfunction, my team used both data (explicit) and our combined insights (tacit) to diagnose the root cause. Formal risk models flagged the possibility; however, our seasoned technicians provided the intuition to spot subtle clues. This aligns with what Rushby proposed: we need to organize our knowledge to *reduce doubt*, blending evidence with expert judgment.

DECISION-MAKING UNDER UNCERTAINTY

Safety decisions are ultimately choices made under uncertainty about outcomes. Classical decision theory frames choices with probabilities and utilities; however, real-world decisions rarely come with neat numbers. Instead, engineers often rely on heuristics and principles. Psychology has taught us that such mental shortcuts can both help and mislead. For example, Kahneman and Tversky's work on heuristics [11] shows that people often overestimate rare disasters if they are vivid in memory (the availability bias). In safety, this means a dramatic recent accident might loom larger in decision-making than statistics warrant, or vice versa, a lack of accidents might engender dangerous complacency. Recognizing these biases is part of epistemic humility.

Epistemology, the study of knowledge, addresses fundamental questions: "What is knowledge?" and "How can claims to knowledge be justified?" Ancient philosophers established the foundation for these inquiries. Plato defined knowledge as the intersection of truth and belief: to truly know something, one must both believe it and have it be true, supported by justification. This classic "justified true belief" model has direct relevance for safety: merely believing that a practice is safe is insufficient unless the belief is supported by evidence, and truth alone does not constitute knowledge without justification. A childhood experience illustrates this principle. While building a small go-kart, the frame appeared sturdy, and it was believed safe to race. When the axle snapped under load, it became clear that belief without proof can fail catastrophically. The incident demonstrates Plato's insight in practice: knowing something is safe requires both belief and verification. Aristotle expanded on these ideas by distinguishing types of knowledge. Ethics and safety, he argued, are rooted not merely in rules but in practical wisdom or phronesis. Knowing the right course of action requires experience and character, not just memorization of procedures. As Aristotle observed, "Practical wisdom cannot be acquired solely by learning general rules. We must also acquire, through practice, those skills that enable us to put our general understanding into practice in ways that are suitable to each occasion." Modern safety practice reflects this principle: workers need judgment to apply procedures effectively in real-world contexts, especially when situations are ambiguous or unpredictable. The interaction between explicit rules, tacit understanding, and collective culture further illustrates this point. Explicit procedures guide action, but they are effective only when reinforced by tacit knowledge and a positive safety culture. For example, a procedure may instruct workers to wear harnesses, but compliance is ensured only when an experienced foreman emphasizes the rule and a culture discourages shortcuts. Similarly, engineering education demonstrates this interplay: students may have extensive formulas and theoretical models, yet practical wisdom is developed by confronting variability and testing assumptions, validating knowledge through experience rather than relying solely on idealized models. Modern epistemology underscores that knowledge is never absolute. Every safety judgment carries epistemic uncertainty, uncertainty about what is truly known. Niklas Möller argues that a robust concept of safety must explicitly account for this uncertainty, defined as "the degree to which we are uncertain of our knowledge of the situation at hand." In practice, this requires asking: How confident are we that a procedure will work? What aspects of this system remain unknown? Overconfidence in data, such as assuming that "since no crashes have occurred, the plant must be safe," can be dangerous. Safety involves not only reducing measurable risk but also addressing epistemic doubt, recognizing blind spots, and expanding the knowledge base through testing and validation. A real-world example illustrates this: at a chemical plant, an experienced operator claimed that a valve was "inherently safe" because it had always failed closed. A more thorough inspection revealed it could fail open under certain conditions, exposing a hidden hazard. This reinforces Aristotle's lesson: practical wisdom requires verifying assumptions rather than relying on routine or hearsay, and it aligns with the principle that reducing epistemic doubt is central to reliable safety knowledge. Safety decision-making is also shaped by values. The precautionary principle reflects a Kantian approach [12] to risk: when unknown hazards exist, err on the side of caution. This principle is not purely data-driven but morally grounded, prioritizing human welfare. In practice, introducing a new chemical process demonstrated this principle: although empirical data suggested safety, ethical consideration required delaying implementation until uncertainties were reduced, ensuring that actions were both effective and morally responsible. Together, these perspectives, Plato's justified true belief, Aristotle's phronesis, epistemic doubt, tacit and collective knowledge, and ethical values, provide a comprehensive philosophical foundation for safety [13]. Rigorous safety decisions depend not only on evidence and rules but also on reflective judgment, experience, cultural reinforcement, and moral responsibility. Integrating theory with practice ensures that safety knowledge is both reliable and ethically sound, guiding decisions in complex, uncertain, and high-stakes environments.

Engineers are trained in formal decision tools, such as fault trees and risk matrices, yet actual choices often rely on phronesis. Aristotle emphasized that ethical principles do not produce algorithmic decisions; likewise, safety ethics frequently require judgment. Determining when to halt production due to ambiguous readings or how aggressively to mitigate unlikely failure modes are not black-and-white questions. Consider a late-night emergency in a chemical unit: an unforeseen fire alarm could have been a false signal or a real fire, while automated data remained inconclusive. Cost–benefit analysis alone, factoring in production loss, proved insufficient given the human stakes. Drawing on experience, tacit knowledge, and the principle of precaution, evacuation was ordered. The alarm was real, averting disaster. This episode illustrates Aristotle's practical wisdom: effective safety decisions often demand acting on reasoned judgment in context rather than relying solely on abstract probability calculations.

Underlying effective safety decisions is an essential intellectual virtue: intellectual humility. Regardless of expertise, knowledge has limits, as Möller emphasizes. This requires listening to dissenting voices and fostering critical debate, echoing Plato's insistence on examining whether beliefs are truly justified and true. Even rational thinkers such as Descartes endorsed civil discourse as a method for checking error. In practice, challenging consensus can reveal hidden risks. For example, a proposed design change to process piping initially seemed unnecessary to most colleagues, yet re-evaluation and testing demonstrated that it eliminated a rare but potentially dangerous pressure wave. This illustrates that belief alone is insufficient; robust safety decisions require a firm basis of reason and evidence to support them.

LEARNING AND EXPERIENCE

Over time, knowledge is revealed to be dynamic, constructed through cycles of action and reflection. Donald Schön's concept of the "reflective practitioner" captures this process, echoing Aristotle's and Dewey's emphasis on learning from experience. Following any safety incident or near-miss, organizations engage in post-event learning: debriefing teams, updating procedures, and internalizing lessons. Each event reshapes both tacit and explicit knowledge. For example, after a fire drill and subsequent review, a company instituted a real-time alarm verification procedure. Official manuals codified the explicit knowledge, while the organizational culture reinforced the prioritization of safety over production, embedding collective knowledge that extends beyond formal rules.

A formative flashback from university illustrates how reflection on knowledge itself shapes learning. In an engineering ethics class, students debated whether blind adherence to regulations constitutes a moral duty. One perspective favored strict rule-following, while another cited Aristotle: "Ethical theory must match its subject matter, good action, and must respect that in this field many generalizations hold only for the most part." This insight, that ethics requires flexibility, proved enduring. Years later, in a project involving a supplier's proposal for a cheaper but slightly weaker material, the rulebook did not forbid its use, yet caution prevailed. Recognizing that regulations are generalizations that may not fit every situation, the material was tested under realistic stress conditions. Practical evaluation revealed a flaw under high humidity. This example demonstrates that effective safety judgment combines rules with contextual assessment, ensuring that procedures are not applied blindly but guided by practical wisdom.

Personal experience also reveals the importance of humility in expertise. Early professional development can foster excessive confidence in formal training, as though technical education provides complete answers. A humbling incident occurred when a seasoned machine operator suggested adjusting a valve setting differently from what engineering calculations recommended. The initial response emphasized data and analytical justification. The operator replied with an insight reminiscent of Polanyi's view of tacit knowledge: sometimes understanding precedes measurement. A compromise was reached by cautiously adjusting the setting while monitoring performance. The adjustment prevented vibration fatigue that the narrow analytical model had overlooked. This episode reinforced a critical lesson: safety decision-making is not purely technical; it is profoundly human. Effective safety practice requires openness to multiple forms of knowledge, including experiential and unarticulated insights. Intellectual humility enables the integration of formal analysis with tacit understanding, strengthening both judgment and outcomes.

BRIDGING THEORY AND PRACTICE

Throughout this journey, a consistent theme emerges: theory and practice must continually inform one another. Engineering education provides formulas and analytical models that are indispensable, yet lived experience, whether through incidents, failures, or near-misses, reveals dimensions of reality that models alone cannot capture. Practical exposure deepens understanding by confronting uncertainty, variability, and consequence. Conversely, real-world practice benefits from the critical discipline of philosophy. By questioning assumptions and demanding justification, philosophical reflection guards against overconfidence and complacency. Aristotle's insight reminds us that rules function as guides rather than infallible decrees; they require contextual judgment and experience to be applied wisely. The integration of theoretical rigor with reflective practice forms the foundation of mature safety decision-making.

In practice, bridging the gap between theory and application often required creating structured forums for critical reflection on site. As part of safety leadership initiatives, regular "philosophy rounds" were introduced, where teams discussed not only how tasks should be performed but also why they were performed in particular ways. These sessions encouraged deeper examination of underlying assumptions and knowledge claims. Philosophical ideas provided useful prompts. Invoking Descartes' method of doubt encouraged engineers to articulate uncertainties explicitly rather than conceal them. Referencing Kant's emphasis on autonomy of understanding reinforced the insight that frontline workers often possess context-specific knowledge that may exceed that of remote managers. Although philosophical sources were not quoted formally, the discussions embodied a reflective spirit: participants critically examined whether their safety beliefs and procedures were genuinely justified.

Another practical approach involved adopting a mindset akin to scientific inquiry. Whenever a new safety measure was proposed, teams were encouraged to ground their conclusions on a firm basis of evidence, consistent with Dewey's conception of reflective inquiry. Rather than accepting a vendor's safety claims at face value, small-scale trials were conducted. Instead of assuming that a checklist captured all potential hazards, simulations and drills were implemented to test underlying assumptions. This disciplined approach to validation gradually became embedded within the safety culture. Over time, team members began to anticipate this reflective lens, strengthening analytical rigor and shared accountability. A sustained culture of inquiry fosters collective learning, ensuring that safety decisions are continuously examined, tested, and refined.

In sum, bridging theory and practice in safety extends far beyond academic reflection. It is expressed in every design review, incident investigation, and informal discussion about risk. Philosophical traditions offer practical guidance: Plato's emphasis on justified belief underscores the need to align conviction with evidence; Descartes' method of doubt encourages systematic questioning; Aristotle's practical wisdom highlights the importance of contextual judgment; and Dewey's pragmatism reinforces learning through inquiry and experience. These perspectives have tangible implications. They require checking beliefs against facts, valuing experiential insight alongside formal analysis, and actively reducing uncertainty wherever possible. The developmental arc of professional growth demonstrates that such lessons are not innate. They are cultivated through cycles of reflection and action, gradually shaping a more mature and responsible approach to safety decision-making.

To conclude, Figure 1.1 presents a conceptual overview of how epistemological theories form the foundation of modern safety knowledge and decision-making. The diagram visually integrates the three major domains discussed throughout the chapter: philosophical origins, core safety principles, and industry practice. On the left, classical and modern philosophers such as Plato, Aristotle, and Dewey provide the epistemological grounding, ranging from justified true belief to practical wisdom (phronesis) to learning through inquiry. These ideas flow into the center of the chart,

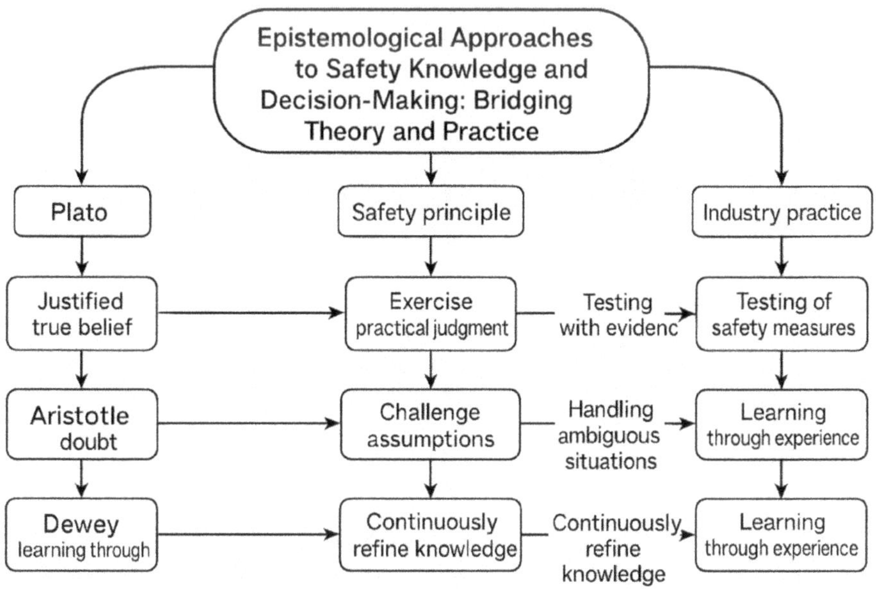

FIGURE 1.1 Epistemological foundations of safety: connections between philosophical concepts, safety judgment, and real-world practice

where they shape essential safety principles such as exercising practical judgment, challenging assumptions, and continuously refining knowledge.

Table 1.1 presents a structured comparison of key epistemological concepts and their practical influence on safety thinking and decision-making. Each row links a major philosophical idea with its corresponding implication for safety practice, supported by illustrative examples drawn from engineering or operational contexts. Together, these comparisons demonstrate how abstract theories of knowledge directly shape real-world safety behavior. The first dimension, "What counts as knowledge?," highlights Plato's conception of knowledge as justified true belief, emphasizing that beliefs about safety must be both accurate and evidence-based. The example of the go-kart axle failure illustrates how untested assumptions can lead to harm when belief is mistaken for knowledge. The second dimension, "Applying rules wisely," draws on Aristotle's concept of phronesis (practical wisdom). This

TABLE 1.1
Epistemological Concepts and Their Implications for Safety Knowledge and Practice

Dimension	Core Idea (Philosophy)	Safety Implication	Illustrative Example / Story
What counts as knowledge?	Plato: Knowledge = Justified True Belief (JTB), belief must be true *and* justified	Don't assume safety; *demonstrate* it with evidence.	Go-kart axle snaps → belief ≠ knowledge without verification (test/calculation).
Applying rules wisely	Aristotle: Phronesis (practical wisdom), good judgment in context, not just rule-following	Procedures need situational judgment; ethics "for the most part" not absolute.	Students pushed beyond textbook cases; foreman emphasis makes harness rules stick.
Challenging assumptions	Descartes: Methodical doubt, question what can be doubted	Run pre-mortems; don't trust "we've always done it this way."	Valve assumed "inherently safe" → failed open under stress test.
Bridging models and data	Kant: Synthesis of rationalism and empiricism, concepts structure experience	Combine models + sensor data; principles guide, evidence applies.	Risk model + plant data to predict failure envelopes.
Learning by inquiry	Dewey: Inquiry builds justified belief	Trial, iterate, update beliefs with reasons; avoid passive acceptance.	Vendor claim tested via small-scale trials before rollout.
What safety knowledge is	Aven: Safety science = facts + conceptual tools	Build both empirical knowledge and frameworks (risk models, principles).	Incident data + safety case structures; define terms, scope, assumptions.

emphasizes that safety procedures alone are insufficient without contextual judgment. The example of students being pushed beyond textbook cases and the foreman reinforcing harness use shows how rules must be interpreted sensibly in real conditions. The third dimension, "Challenging assumptions," connects to Descartes' methodical doubt, which promotes questioning what is routinely accepted. The valve failure episode demonstrates the danger of trusting historical patterns without verification and the need for critical skepticism in safety work. The fourth dimension, "Bridging models and data," reflects Kant's synthesis of rationalism and empiricism, where conceptual models guide understanding but must be grounded in observable data. This is illustrated through risk modeling validated by plant sensor information, showing how safety knowledge emerges from both theory and empirical evidence. The fifth dimension, "Learning by inquiry," is based on Dewey's view that inquiry is essential for forming justified beliefs. In safety practice, this translates into trial, iteration, and continuous improvement, as shown by testing vendor claims through small-scale trials before implementation. Finally, the sixth dimension, "What safety knowledge is," aligns with Aven's definition of safety science as a combination of empirical knowledge and conceptual tools. This includes formal risk models, principles, and structured methods. The example of incident data informing safety cases demonstrates how factual evidence and conceptual frameworks jointly support robust safety reasoning. Overall, Table 1.1 synthesizes philosophical foundations with practical safety lessons, illustrating that safety is not merely procedural but deeply epistemological: it depends on how knowledge is justified, questioned, verified, and applied in real contexts.

CONCLUSION

Safety is ultimately grounded in knowledge: knowing which hazards exist, understanding how systems behave, and discerning how to act appropriately. This chapter has emphasized that examining how knowledge is formed and justified is as important as the content of that knowledge itself. Philosophical traditions offer essential lenses for this inquiry. From early experimentation to complex crisis management, professional development demonstrates that every safety lesson carries an epistemological dimension. Plato's insight warns that unexamined belief can lead to harm. Aristotle shows that sound judgment must be cultivated through experience and practice. Kant highlights that knowledge emerges through the interplay of empirical evidence and rational reflection. Dewey reinforces that knowledge advances through deliberate inquiry and testing. Together, these perspectives affirm that effective safety practice depends not only on technical competence but also on reflective understanding of how knowledge is constructed, questioned, and refined.

By bringing these ideas into the workshop and the boardroom, safety practice becomes more rigorous and more reflective. Questioning assumptions, inspired by Descartes' method of doubt, strengthens critical evaluation. Accepting and actively reducing uncertainty, as emphasized by Möller, enhances the robustness of safety judgments. Integrating technical evidence with human insight, consistent with Rushby's perspective, ensures that formal analysis is complemented by contextual understanding. Professional development over time reveals that each formative episode represents a microcosm of this broader learning process, an interplay between evidence and belief, between formal rules and practical judgment. Such integration transforms safety from a procedural obligation into an intellectually grounded and ethically responsible discipline.

Ultimately, bridging theory and practice requires recognizing that safety knowledge is not merely a collection of facts but an ongoing dialogue between understanding and reality. Philosophy enriches this dialogue by urging precise definitions, careful justification of methods, and alignment of action with well-grounded reasons. In industrial contexts, where errors can carry severe consequences, such intellectual discipline is indispensable. Adopting an epistemological perspective, continually asking "How is this known?," transforms routine compliance into deliberate stewardship. It shifts safety from mechanical adherence to procedures toward reflective responsibility. Professional growth, therefore, depends not only on mastering formulas and protocols but also on cultivating the wisdom preserved in philosophical thought. Learning from philosophy is not an abstract exercise for safety practice; it is foundational to making sound, responsible decisions in environments where certainty is rare and consequences are real.

REFERENCES

1. Yazdi, M., 2025. From harm to harmony: What philosophers teach us about safety. In Yazdi, M. (eds). *Safety-Centric Operations Research: Innovations and Integrative Approaches*. Studies in Systems, Decision and Control, Springer. https://doi.org/10.1007/978-3-031-82934-5_1
2. Pritchard, D., 2013. *What is This Thing Called Knowledge?*. Routledge.
3. Barnes, J., 1980. Aristotle and the methods of ethics. *Revue internationale de philosophie*, 34, pp.490–511.
4. Perinetti, D., 2014. Ways to certainty. In *The Routledge Companion to Eighteenth Century Philosophy* (pp. 265–293). Routledge.
5. Dekker, S. and Conklin, T., 2014. *Safety Differently*. London: CRC Press.
6. Vanzo, A., 2016. Empiricism and rationalism in nineteenth-century histories of philosophy. *Journal of the History of Ideas*, 77(2), pp.253–282.
7. Dicker, G., 1972. John Dewey on the object of knowledge. *Transactions of the Charles S. Peirce Society*, 8(3), pp.152–166.
8. Aven, T., 2025. *A History of Risk Science: Concepts and Principles of Risk Analysis.*
9. Möller, N., 2012. The concepts of risk and safety. In *Handbook of Risk Theory: Epistemology, Decision Theory, Ethics, and Social Implications of Risk*, 1, pp.55–85.
10. Rushby, J., 2013, September. Logic and epistemology in safety cases. In *International Conference on Computer Safety, Reliability, and Security* (pp. 1–7). Berlin, Heidelberg: Springer Berlin Heidelberg.
11. Tversky, A. and Kahneman, D., 1973. Availability: A heuristic for judging frequency and probability. *Cognitive Psychology*, 5(2), pp.207–232.
12. Van Der Sluijs, J.P., Kaiser, M., Beder, S., Hosle, V., Kemelmajer de Carlucci, A. and Kinzig, A., 2005. The precautionary principle.
13. Eikeland, O., 2008. *The Ways of Aristotle: Aristotelian Phronesis, Aristotelian Philosophy of Dialogue, and Action Research* (Vol. 5). Peter Lang. Bern, Switzerland

2 Governing Risk in the Digital Age

Universal Principles Across Finance, Platforms, and Safety

INTRODUCTION

Risk is an inescapable feature of modern life, woven into activities ranging from routine online purchases to high-stakes finance. As the sociologist Ulrich Beck famously observed, ours is a "risk society" organized around managing the hazards of its own making [1]. The previous chapters of this book have explored philosophical perspectives on safety and risk in various contexts, from the gig economy of food delivery to the virtual world of cryptocurrency, each revealing unique dangers and moral dilemmas. This final chapter seeks to combine all these themes into a unified and practical framework for understanding risk and safety across domains. By examining common patterns and interdisciplinary linkages, we aim to move "from harm to harmony": translating insights into strategies that reduce harm and foster more harmonious (i.e., safe, fair, and trustworthy) outcomes in socio-technical systems.

- **The Challenge of Diverse Risks**: The domains under review appear disparate. Consider a few examples: food delivery platforms promise convenience but often at the cost of precarious labor and accident risk for couriers; auto lending expands access to mobility but can trap borrowers in debt or expose lenders to defaults; gambling offers entertainment and revenue yet spawns addiction and financial ruin for some; cryptocurrency enables new financial innovations but is rife with scams and wild market swings; insurance provides safety nets but faces crises when risks are mispriced or underregulated; auctions (from eBay to Google's ad exchanges) create efficient markets but can be gamed by insiders or plagued by the "winner's curse"; loyalty programs reward consumers but also "hack our brains" to encourage overspending and enable data exploitation; and fake online reviews distort reputations, undermining trust in digital marketplaces. Despite surface differences, these scenarios share underlying questions: How do we define acceptable risk? Who bears the harm when things go wrong? How can systems be designed or regulated to protect people without stifling innovation or freedom?

DOI: 10.1201/9781003634720-2

- **Philosophical Meets Practical**: Philosophers have long debated such questions. For example, the ethics of risk involves balancing potential benefits against harms, under uncertainty, while respecting rights and fairness [2,3]. A recurring theme is the distribution of risk, ensuring no group is unjustly burdened by dangers or excluded from benefits. The *precautionary principle* argues for erring on the side of safety in the face of unknown hazards, whereas others advocate for calculated risk, taking it as a driver of progress [4]. These abstract principles become concrete when examining real cases: e.g., should a food delivery company prioritize speed (customer satisfaction) or courier safety (risk reduction)? How much freedom should individuals have to gamble or invest in unregulated crypto, knowing the potential for self-harm? When does *caveat emptor* ("buyer beware") no longer suffice, requiring protective intervention?

In this introduction, we set the stage by highlighting the need for a holistic approach. The complexity of contemporary risk demands bridging disciplinary silos. Engineering provides technical safety mechanisms; economics offers risk modeling and incentive design; psychology reveals how cognitive biases (like impulsivity in gambling or trust in reviews) can amplify risk; ethics and law guide—what should be permissible or prohibited. We aim to integrate these perspectives. The sections that follow will delve into cross-cutting themes and illustrate them with global examples. By the end, we propose a practical framework, distilled into guiding principles and strategies, that can help engineers, regulators, and practitioners transform insights into action.

A PANORAMA OF RISKS ACROSS DIVERSE DOMAINS

To build a unified framework, we first survey the landscape of risks in the key domains featured in this volume. Each domain has been studied in earlier chapters; here we summarize their salient risks and real-world manifestations:

- **Gig Economy and Food Delivery**: The rapid growth of app-based food delivery has introduced significant occupational safety risks. Couriers (often informal workers) face traffic accidents, time pressure, and exposure to weather and urban hazards. A study in Malaysia recorded 1,242 road crashes involving food delivery riders between 2018 and 2022, including 112 fatalities and 82 serious injuries [5]. In Vietnam, over half of surveyed delivery motorcyclists had been in a crash in just the past year. Similar trends are reported worldwide as millions of riders race to meet algorithmically set deadlines. The gig work model often shifts risk onto workers: lacking traditional employment protections, many couriers must personally absorb medical costs or lost income after accidents. This raises questions of fairness and responsibility; how should platforms, restaurants, customers, and governments share the duty of keeping delivery workers safe? Case studies show some platforms experimenting with safety training, providing helmets, or adjusting dispatch algorithms to discourage reckless speed. Regulators in some jurisdictions have also stepped in (e.g., setting minimum pay per delivery or mandating insurance). Yet, trade-offs emerge between efficiency and safety and between workers' autonomy and paternalistic rules. Gig economy risks thus exemplify the tension between innovation (the convenience of on-demand delivery) and harm (injuries and exploitation) that a harmonious approach must reconcile.

- **Consumer Finance and Car Loans**: Automobile loans are a double-edged sword, enabling car ownership (often essential for work and daily life), but also exposing households and lenders to financial risk. In recent years, auto loan delinquencies have climbed markedly, signaling growing stress. In the United States, the portion of auto loan balances over 30 days past due reached 3.8% by mid-2024, the highest level since 2010 [6]. Analysts note that what was once considered a safe lending segment has become one of the riskiest, with auto loans now experiencing higher default rates than credit cards or mortgages. Several factors drive this: record high car prices (the average new car costs over $50,000) and rising interest rates have pushed monthly payments to unprecedented levels; one in five new loans in the United States carries a payment above $1,000. Borrowers, especially subprime customers, increasingly struggle to keep up, and repossession of vehicles (a harsh outcome) has risen. The social implications are serious: default can mean loss of mobility and employment for individuals, while at scale these bad loans can pose systemic risks reminiscent of the subprime mortgage crisis. Predatory lending practices (e.g., "yo-yo" financing traps or hidden fees) exacerbate harm. Regulators like the US Consumer Financial Protection Bureau (CFPB) have tried to crack down on abusive auto lending, but enforcement has been uneven [7]. Car loan risks highlight the need for *responsible finance*: aligning lenders' profit motives with borrowers' long-term well-being through measures like ability-to-pay assessments, fair interest rate caps, and offering insurance or flexibility to handle life setbacks. These themes resonate with broader questions of economic justice and the ethics of debt.
- **Gambling and Gaming**: Legalized gambling has proliferated globally, from casinos and lotteries to online betting apps, bringing entertainment to many but severe addiction and public health harms to a significant minority. A comprehensive international commission in 2024 (in *The Lancet Public Health*) warned that net losses by gamblers worldwide could exceed US$700 billion by 2028 [8], given current growth trends. Over 80% of countries now permit some form of legal gambling, and nearly all others are reached by unregulated online gambling across borders. The societal costs are multifaceted: an estimated 1.2% of the world's adult population meets criteria for gambling disorder, and around 5.5% of women and 11.9% of men have experienced harm from gambling (e.g., financial stress, mental health issues). Young people are particularly at risk; more than one in four teenage gamblers is at risk of developing a serious problem.

The ripple effects mean that for each problem gambler, an average of six other people (family, friends) suffer consequences such as broken relationships or economic hardship. Notably, the normalization and digitalization of gambling are key drivers of harm: smartphone betting apps, in-game microtransactions (akin to gambling in video games), and ubiquitous sports betting advertising create constant temptation. Ethical concerns arise about exploiting cognitive biases, for instance, "dark nudges" in electronic gambling machines that mislead players about their chances or loyalty rewards in casinos that encourage longer play despite losses. Regulators have begun shifting from viewing gambling purely as an individual vice to treating it as a public health issue. Examples include mandatory loss limits, self-exclusion programs, restrictions on advertising, and banning credit card use for betting (as done in the UK and Australia). The Lancet Commission urged that health and well-being be prioritized over tax revenues, calling for stronger international cooperation and independent national regulators to curb industry excesses. Gambling encapsulates the moral question of how society should handle consensual yet harmful risk; it tests the balance between personal freedom, industry profits, and the duty to prevent harm.

- **Cryptocurrency and Digital Assets**: The crypto boom of the past decade has created new wealth and technological innovation (e.g., blockchain applications) but also introduced profound market and security risks due to its largely unregulated nature. Cryptocurrencies like Bitcoin are notoriously volatile; prices can swing wildly, leading to massive gains for some and devastating losses for others in short periods. Beyond volatility, the crypto ecosystem has been plagued by fraud and scams that exploit naive investors. Many of these scams are "investment" scams; for example, bogus crypto trading opportunities are often tied to elaborate online romance scams (so-called "pig butchering" schemes) that bilk victims of life savings. High-profile collapses have underscored the risks: the 2022 failure of FTX, a major crypto exchange, revealed how the lack of transparency and basic safeguards (like audited reserves or consumer protections) can result in billions of dollars vaporizing overnight, with cascading effects on trust across the industry. Crypto also poses regulatory dilemmas globally. Some countries (El Salvador) have embraced it as legal tender, while others (China) have banned certain crypto activities, and most are scrambling to fit crypto into existing financial laws or draft new ones. Key concerns include money laundering, hacking/theft (hundreds of millions have been stolen from exchanges and DeFi platforms by hackers), market manipulation, and the absence of recourse for victims (transactions are irreversible by design). The crypto domain vividly illustrates the interplay of technological risk (complex code and protocols), market risk (speculation bubbles), and regulatory risk (jurisdictional arbitrage, where scammers migrate to lax regimes). It raises the question: How can we harness innovation (e.g., the efficiency of decentralized finance or the transparency of blockchain) while containing the harm? Potential paths involve international cooperation to set minimum standards, encouraging self-regulation and best practices in the industry, and developing better consumer education (so individuals can spot red flags and secure their assets). Crypto's story is a microcosm of managing risk in the face of rapid innovation, a theme likely to recur with AI and other emerging tech.

- **Insurance and Systemic Safety Nets**: The insurance sector is traditionally society's mechanism for moving from harm to harmony. By pooling risk, it spreads losses and cushions individuals from catastrophic shocks. However, insurance itself faces challenges when underlying risks shift or concentrate. A salient example is catastrophic climate-related events. The frequency and severity of natural disasters (hurricanes, floods, wildfires) have increased, leading to soaring payouts. Thanks to strong insurance penetration in some markets, a larger share of these losses was covered than ever before; globally about 62% of disaster losses were insured in 2025 [9], a record high (leaving 38% uninsured). Yet this masks huge disparities: outside high-income countries, most losses remain uninsured. In emerging economies, over 90% of disaster losses can be uninsured, reflecting an enormous insurance protection gap. When people or governments lack insurance, the impact of harm is much more severe and long-lasting, undermining resilience. Insurance markets themselves can become unstable if risks are not well managed; e.g., if a disaster beyond modeled expectations strikes, insurers may go bankrupt (as seen in the case of certain coastal homeowners' markets facing repeated hurricanes). There is also an ethical dimension: insurance can lead to moral hazard, where being insured might prompt riskier behavior (e.g., a driver being less careful because they have full coverage). Likewise, issues of fairness and access arise; poorer communities often face higher premiums or exclusion, contributing to inequality in safety nets. Recent global discussions emphasize that insurers must shift from passively paying claims to actively investing in risk reduction, for example, offering premium discounts for disaster-proof home upgrades or partnering with governments on infrastructure improvements. In summary, the insurance domain teaches the importance of collective approaches to risk and the perils when risk exceeds our governance capacity. It urges a proactive approach: not just transferring risk but *reducing* it.
- **Auctions and Market Mechanisms**: Auctions are employed in many fields (art, commodities, spectrum sales, online advertising) to allocate resources efficiently. However, poorly designed or manipulated auctions can cause participants harm or distort markets. One classical auction pitfall is the "winner's curse"; the winner may overpay relative to the item's intrinsic value, a risk especially in common value auctions (where the item's value is uncertain). For instance, during the 2000s, some 3G telecom spectrum auctions saw companies pay exorbitant prices that impaired their finances for years, a market-level harm stemming from auction dynamics. Another concern is collusion and manipulation. There have been cases of bidders covertly agreeing to suppress bids to win at lower prices, defeating

the auction's competitive purpose (e.g., documented collusion in some Treasury debt auctions). In the digital realm, auctions underpin the real-time bidding for online advertisements. Here, information asymmetry and platform power create risks: recent antitrust investigations revealed that Google, which operates the largest ad exchange, ran internal programs (with code names like "Project Bernanke") that manipulated ad auctions to its advantage, allegedly skimming revenue from publishers and advertisers. According to the US Department of Justice, Google's ad exchange secretly gave itself a positional edge and used data from across its ecosystem to win bids at optimal prices, undermining fair market outcomes. Such conduct not only harms advertisers and publishers (who earn less or pay more than they would in a transparent market) but also erodes trust in the auction mechanism as a fair allocator. Beyond monopoly issues, auctions can entail systemic risk when they fail: a notable example was the 2008 crash of the Auction Rate Securities (ARS) market; auctions for certain bonds froze, leaving investors unable to access funds and issuers facing exorbitant rates. The ARS failure highlighted the danger of assuming liquidity and safety in a complex auction without proper safeguards; it led to settlements and reforms as regulators stepped in post hoc. The lesson from auctions is that market design matters: clear rules, transparency, and oversight are needed to prevent exploitation. In a harmonious risk framework, auction designers (or algorithm programmers) carry the responsibility to foresee failure modes or abuses, and regulators must remain vigilant where market incentives alone will not ensure integrity.

- **Loyalty Programs and Consumer Behavior Engineering**: Customer loyalty programs, from airline miles and hotel points to retailer rewards apps, are generally seen as benign marketing tools. Yet, they can introduce subtle risks and ethical issues by exploiting behavioral biases and even undermining consumer interests. Companies have learned that points, tiers, and personalized offers can "hook" customers by triggering psychological rewards (dopamine hits) like gambling, fostering habit formation, and even addiction-like behavior in shopping. For example, credit card reward programs have been shown to encourage overspending as consumers chase points, effectively hacking the brain's reward system [10]. Additionally, loyalty programs often involve extensive data collection on purchase behavior, raising privacy concerns and enabling discriminatory pricing. A 2025 investigative report titled *"The Loyalty Trap"* [11,12] found that many companies quietly devalued rewards or imposed hidden fees, leveraging the data from loyalty schemes to charge loyal customers more over time. One case saw Grubhub (a food delivery firm) accused of promising "free delivery" for members but then adding junk fees, a practice so deceptive that the US Federal Trade Commission and the state of Illinois sued the company in 2024. Another aspect is loyalty fraud and security: hackers target loyalty point accounts (which can be like virtual currency); in fact, airline miles and hotel points theft has been on the rise as an underground criminal economy. For consumers, a risk is also "loyalty lock-in," feeling stuck with one provider to not lose accrued rewards, even if it is not the best choice (a form of switching cost that can be exploited by firms). From a harm perspective, loyalty programs blur the line between incentivizing and manipulative. Regulators have begun to scrutinize "dark patterns" in subscription services (e.g., making it hard to cancel a loyalty-based subscription); indeed, the Grubhub case also involved allegations of creating obstacles for users trying to cancel their paid memberships. The overarching point is that design choices in software and programs can create risks: here, the risk is exploitation of consumers' cognitive vulnerabilities and erosion of consumer surplus. An ethical, harmonious approach would urge transparency (clear terms and real savings), fairness (rewards that truly benefit loyal customers rather than mislead), and respect for autonomy (easy opt-out, no addictive gimmicks).

- **Fake Reviews and Online Misinformation**: In digital marketplaces and social media, information integrity is crucial for safety and trust. Fake reviews, fabricated or paid for ratings and testimonials, have emerged as a significant risk, misleading consumers and harming honest businesses. Studies have shown that over 80% of shoppers consult online reviews before buying a product, meaning manipulation of this content can skew a huge volume of transactions. Fake positive reviews can dupe people into buying low-quality or unsafe products, while fake negative reviews can unfairly destroy a business's reputation. The economic harm is tangible: one analysis by the UK Competition and Markets Authority estimated that billions of pounds of consumer spending are influenced by potentially inauthentic reviews each year (CMA, 2020). Moreover, fake reviews erode trust in the whole platform. If users come to doubt the ratings, the mechanism of crowdsourced quality assurance breaks down, a classic example of how deceit by some can impose risk (in this case, decision risk and monetary loss) on the public. Governments have started to respond. The rule even targets company insiders posting reviews without disclosure and the burgeoning problem of AI-generated reviews. This regulatory stance recognizes that fake reviews "pollute the marketplace and divert business away from honest competitors." Ensuring authenticity online is thus a safety issue in a broad sense; it protects consumers from fraud and companies from defamation. Tackling it requires both technological solutions (algorithms to detect suspicious patterns, such as clusters of reviews from the same IP) and legal deterrence (penalties to raise the cost of cheating). The fake reviews problem ties back to a fundamental principle: transparency and truthfulness are the bedrock of any safe, fair system. Where information asymmetry or deception exists, the risk of harm rises, be it financial loss or even physical harm (imagine a fake positive review on a dangerous product). This domain teaches that *information risk* is as important as material risk, especially in the digital age.

Each of these domains, and the others alluded to, provides a piece of the puzzle. They show that risk is multidimensional: physical safety, financial security, psychological well-being, trust and reputation, cybersecurity, and ethical governance are all at play. Despite the differences, we can already see recurring motifs. For instance, information asymmetry appears in several forms (loan contracts that borrowers do not fully understand, hidden odds in gambling, opaque algorithms in auctions, undisclosed incentives in loyalty programs, or fake information in reviews). Similarly, misaligned incentives are a common culprit (if a company profits while users incur harm, as seen with some casinos, crypto exchanges, or platforms that externalize risks to gig workers, problems ensue). The human factor, whether it is cognitive biases like impatience and overconfidence, or social factors like peer influence, also recurs, reminding us that risk systems must account for real human behavior, not just idealized rational actors.

COMMON THREADS AND INTERDISCIPLINARY LINKAGES

Across the above domains, several interdisciplinary themes emerge. Identifying these can help in formulating a unified framework. Here we discuss the key common threads, linking theory to practice:

Information Asymmetry and Transparency

Many risks are exacerbated by one party knowing more than another. In car loans, lenders traditionally knew more about credit risk modeling than borrowers, but borrowers know more about their intent to repay. In auctions, the auctioneer or a dominant platform might have insider information (as with Google's ad exchange). In online shopping, sellers can fake reviews to create a false image, and consumers lack the means to verify claims. Transparency is the antidote to this asymmetry. For instance, financial regulators push for clear disclosure of loan terms (truth in lending laws) so borrowers understand risks like variable interest rates or penalties. In online marketplaces, platforms like Amazon and TripAdvisor have begun marking "Verified Purchase" on reviews to signal authenticity and deploying AI to filter out likely fakes. Transparency builds trust: consider how the blockchain technology underlying cryptocurrencies is valuable for transparency (an open ledger of transactions), yet ironically the crypto industry has suffered from opaqueness in business operations. A famous example was the collapse of Mt Gox exchange in 2014 and FTX in 2022 [13], in both cases, a lack of transparent accounting concealed mismanagement until too late. The principle here is "sunlight as disinfectant": more open information flows allow risks to be identified and addressed before they compound. Philosophically, this aligns with Kantian ethics [14] about honesty and treating others not merely as means; hiding information that could foreseeably harm others is unethical. Practically, regulators increasingly mandate transparency (e.g., food delivery platforms forced to show how prices/fees are split with restaurants and drivers or gambling operators required to show odds and payout rates). A unified framework would place transparency front and center as a preventive tool, reducing harm by empowering all stakeholders with knowledge.

Incentive Alignment and Moral Hazard

In many risk situations, the root cause of harm is that the incentives of different actors are misaligned. Take insurance: if an insured individual is fully protected from loss, they might take riskier actions (classic moral hazard). Similarly, if a food delivery platform's algorithm rewards faster deliveries without regard to safety, couriers are incentivized to take dangerous shortcuts. If an online retailer's revenue increases with each positive review (even fake ones), they are incentivized to solicit or buy fake reviews unless penalized. Aligning incentives means structuring the system so that what benefits one party also benefits others and the broader system. This often involves internalizing externalities: for example, some cities have considered making ride-sharing companies pay into funds that cover accident costs for drivers, thereby incentivizing the companies to invest in safety training to reduce those costs. In finance, lenders retain skin in the game by keeping a portion of loans they originate rather than selling them all; this alignment (used in reforms after the 2008 crisis) means they suffer losses if loans default, motivating prudent lending. In gambling, alignment is trickier because the house profits from player losses; thus, government intervention (taxes, mandatory contributions to treatment programs, profit limits) might be used to align casino profits with player well-being to some extent. The idea of "safety by design" in engineering can be seen as incentive alignment between designers and users: for instance, a car's seatbelt reminder annoys the driver until they buckle, aligning the car's immediate "interest" (stopping the alert) with the driver's safety. From a philosophical perspective, alignment connects to utilitarian thinking, designing institutions that convert individual rational actions into collective optimal outcomes. A unified framework emphasizes identifying where misalignments exist and correcting them via policy (fines, rewards, regulations) or design (protocols, defaults) so that actors who create risk also bear consequences and actors who reduce risk are rewarded. Table 2.1 later in this chapter illustrates some misalignments and how they can be fixed.

Behavioral Dynamics and Ethical Design

Human behavior often deviates from the "rational actor" model, which can amplify risk. Behavioral economics and psychology provide insight here: cognitive biases (present bias, overconfidence, herd mentality, etc.) are common across domains. Gamblers may fall prey to the "gambler's fallacy" (believing a win is due after a string of losses), leading them to persist in harmful betting. Borrowers may be overly optimistic about future income, underestimating the risk of default. Shoppers place outsize trust in a five-star rating average without reading details (anchoring bias), making fake reviews effective. Recognizing these patterns allows for ethical design interventions. For example, many jurisdictions require a "cooling off" period for gamblers who win big, to counteract the tendency to immediately reinvest winnings

TABLE 2.1
Key Risk Factors and Mitigation Strategies Across Selected Domains

Domain	Key Risks/Harms	Current/Potential Mitigation Strategies
Food Delivery (Gig Economy)	Courier accidents and injuries (traffic risk) Income instability and lack of worker protections Pressure to violate traffic laws for speed	Safety training and equipment provided by platforms Dynamic dispatch algorithms that prioritize safety (e.g., not penalizing slightly slower deliveries) Regulation: minimum wage guarantees, insurance mandates for gig workers Community-based programs (e.g., ride, sharing cooperatives) to empower workers
Auto Loans (Consumer Finance)	Borrower over indebtedness and default Predatory lending and hidden fees Repossession leading to loss of essential asset	Rigorous credit assessments and caps on loan, to, value ratios Transparent disclosure of terms (truth, in, lending laws) Regulatory oversight targeting predatory practices (CFPB enforcement, interest rate caps) Financial literacy programs to improve borrower decision-making
Gambling	Addiction and problem gambling (health harms) Financial ruin, debt, bankruptcy Social harms (family neglect, crime to finance gambling)	Self-exclusion and limit, setting tools for gamblers (often mandated by law) Restrictions on advertising, especially targeting vulnerable groups (e.g., banning ads during sports events for underage viewers) On-site interventions: reality checks (pop-ups after continuous play), mandatory breaks Treatment and counseling services funded by gambling taxes (public health approach) Strong regulation: independent gambling regulators, monitoring of industry practices (as recommended by Lancet Commission)
Cryptocurrency	Fraud scams (Ponzi schemes, fake ICOs) Exchange hacks and theft of assets Extreme price volatility causing investor losses Lack of recourse (no refunds or insurance)	Regulatory frameworks to classify and oversee crypto assets (e.g., anti-fraud provisions similar to securities law) Industry self-regulation: exchanges implementing proof, of, reserves, third-party audits Improved cybersecurity standards for crypto platforms, multi-signature wallets for security Consumer education campaigns about scam red flags and volatility risks Development of insured custody services for crypto (to protect against hacks)

(*Continued*)

TABLE 2.1 (Continued)
Key Risk Factors and Mitigation Strategies Across Selected Domains

Domain	Key Risks/Harms	Current/Potential Mitigation Strategies
Insurance	Protection gaps (large % of population or assets uninsured, especially for disasters) Insolvency risk if catastrophes exceed insurers' capital Moral hazard (risky behavior by insured parties)	Public, private insurance partnerships (e.g., government backstops for extreme events, microinsurance for low-income groups) Risk-based premiums that incentivize risk reduction (e.g., discounts for safer construction) Catastrophe bonds and reinsurance to spread extreme risks globally Data-driven risk modeling (cat models) to better price and prepare for emerging risks (pandemic insurance, cyber insurance) Policy clauses that encourage prudent behavior (deductibles, co-payments to retain some risk with insured)
Auctions and Market Design	Winner's curse (overpayment) leading to losses Collusion or manipulation undermining fairness Auction failures causing market liquidity crises	Careful auction design: use of second-price auctions or reserve prices to mitigate winner's curse Anti-collusion monitoring and antitrust enforcement in auction markets (e.g., penalties for bid-rigging cartels) Transparency in auction process (publishing auction rules, independent oversight of platform, run auctions) Robust contingency planning: e.g., for financial auctions, having a market, maker of last resort, or temporary halt rules if auctions malfunction (to prevent freezes like ARS crisis)
Loyalty Programs	Consumer exploitation via complex point systems (value dilution of points, hidden fees) Privacy intrusion (extensive tracking of purchases) Encouraging excessive spending or unhealthy habits (e.g., casino VIP programs for high rollers)	Transparency requirements: clear disclosure of point value, expiration, and any fees (truth in rewards guidelines) Privacy regulations (e.g., GDPR) limiting data use , consumers can opt out of data sharing without losing basic benefits Ethical design of reward structures (focusing on true loyalty vs addictive "gamification") Regulators scrutinizing unfair or deceptive practices (as in the Grubhub case) and allowing easy exit for consumers (one-click unsubscribe/cancel)

(Continued)

TABLE 2.1 (Continued)
Key Risk Factors and Mitigation Strategies Across Selected Domains

Domain	Key Risks/Harms	Current/Potential Mitigation Strategies
Online Reviews and Reputation	Fake positive reviews misleading consumers Malicious fake negative reviews harming businesses General erosion of trust in online information	Platform measures: verified reviews, AI detection of review farms, user reporting systems Legal measures: making the creation or sale of fake reviews illegal (e.g., FTC rule; EU Unfair Commercial Practices directive) Penalties for businesses caught using fake reviews (fines, delisting from platforms) to deter the practice Consumer education: encouraging skepticism of overly perfect ratings, looking for verified buyer tags, etc. Development of third-party services that certify businesses (e.g., Trustmark certifications) as a supplement to review systems

in further play. Online platforms are experimenting with nudges for good: warning labels on products with suspiciously positive reviews, or pop-up reminders "You have written 10 reviews today, please ensure they reflect genuine experiences" to deter fake review farms. However, there is also the dark side of design, exploitative practices (sometimes called "sludge" or dark patterns). Loyalty programs exploiting the endowment effect (people value points more once earned, so they avoid letting them expire even if it means spending more) or social media apps exploiting intermittent rewards to keep users engaged exemplify design that amplifies risk for profit. Ethical design would flip this, using understanding of behavior to reduce risk: e.g., apps could allow self-imposed limits (some fintech apps let users cap their daily spending or investing to prevent rash decisions), or casinos could redesign slot machines to be less immersive (slower spins, clear odds displayed). Engineers and designers carry a responsibility akin to doctors under "do no harm"; when you know a design element might trigger addictive or reckless behavior, an ethical framework suggests mitigating it. Interdisciplinary collaboration is valuable: psychologists informing product design, ethicists on AI ethics boards, etc., to foresee and preempt harmful outcomes. Ultimately, treating users with respect for their autonomy and welfare, not as mere targets to exploit, builds long-term trust and reduces systemic risk (a user base that does not feel cheated or harmed is less likely to spark backlash or require heavy regulation).

Interconnectedness and Systemic Risk

Another commonality is that risks are often interconnected across domains and borders, forming complex systems. A failure in one component can cascade. For instance, a major insurer's collapse can affect car loans (if auto insurers fail, car lending may freeze since cars cannot be insured properly). Cryptocurrency markets interact with gambling; indeed, crypto trading has been likened to speculative betting, and some problem gamblers migrate to high-risk trading apps. Fake reviews and fraud in e-commerce can reduce trust in online transactions, which can spill over into reluctance to engage with fintech innovations. The COVID-19 pandemic was a prime example of systemic risk: it simultaneously affected health (safety risk) [15], the economy (financial risk, e.g., car loan defaults spiked after job losses), and increased reliance on delivery and online services (amplifying those domain risks). The concept of systemic risk teaches that siloed management is insufficient. We need a holistic view: consider how climate change is now understood not just as an environmental issue, but a financial stability issue (central banks conduct climate stress tests on insurers and banks) and a social issue (affecting migration, inequality). Solutions therefore must be coordinated. This has led to frameworks like "One Health" (connecting human, animal, and environmental health) in safety, and macroprudential regulation in finance (regulators looking at the whole system, not just individual firms). Engineers, too, have embraced systems thinking, safety engineering now examines not just a component but the entire socio-technical system (for instance, aviation safety looks beyond aircraft design to pilot training, air traffic control, maintenance routines, etc., as all parts interact to prevent accidents). For the domains we covered, this means, for example, collaborative governance: tech platforms working with financial regulators to spot crypto scams or healthcare providers teaming with gambling regulators to address gaming addiction as a mental health issue, not merely a moral failing. Philosophically, this aligns with concepts of collective responsibility and the idea that complex problems require joined-up responses. An implication for our unified framework is the emphasis on *communication and data sharing across sectors*. If insurers, banks, tech companies, and governments share relevant risk data (respecting privacy), they can identify emerging risks faster. A real-world illustration is how several governments now share information on crypto exchange malpractices and coordinate enforcement so scammers cannot simply hop jurisdictions, a nascent but necessary global effort.

Regulatory Foresight and Adaptability

In virtually every domain above, regulation lags innovation; often, risks are realized and harms occur before rules catch up. Part of the issue is foresight. Traditional regulatory approaches can be reactive and siloed (e.g., gambling regulated separately from online gaming, even if they converge). There is a need for anticipatory governance, trying to predict and shape emerging risks before they cause harm. The rapid institution of the FTC's anti-fake review rule in 2024, ahead of an expected explosion of AI-generated reviews, is a positive example of foresight: regulators observed AI trends and moved to bolster truth in advertising rules proactively. Adaptability is equally important: risk environments change, as do social expectations. For instance, what was considered an acceptable level of road fatalities decades ago is no longer tolerated; this drove regulators to steadily raise vehicle safety standards and mandate features like airbags. In finance, when new products like Auction Rate Securities emerged, regulators initially treated them under old frameworks, but their unique auction liquidity risk was not fully appreciated until the crash forced adaptation. A unified risk framework champions agile regulation, rules that are principle-based and technology-neutral where possible, enabling adaptation. One trend is regulatory sandboxes (used in fintech, for example), where innovators work with regulators in a controlled setting to pilot new models (like a new lending platform or crypto service) while monitoring risks closely. This way, rules can be iteratively developed alongside innovation. Another approach is co-regulation and industry standards: industries often understand their tech better, so encouraging them to set codes of conduct (if there is oversight to ensure they are not lax) can fill gaps faster than formal laws. Importantly, the involvement of ethicists and philosophers in policy discussions (e.g., AI ethics boards, bioethics in healthcare policy) is on the rise, reflecting that questions of risk inevitably involve values and ethical choices, not just technical calculations. We see this in debates about *autonomous vehicles*: regulators must decide how safe is "safe enough" to deploy self-driving cars (a moral threshold, not just technical). Or in climate risk: how to weigh economic growth against the risk of ecological harm involves normative judgments. Therefore, integrating ethical reasoning into risk governance, essentially blending philosophical perspectives with empirical evidence, is key for sound foresight. Our framework thus calls for regulators to be not only rule enforcers but also *risk strategists*, constantly scanning for new perils (cybersecurity threats, AI manipulation, etc.) and updating approaches in dialogue with stakeholders.

These common threads demonstrate that, whether dealing with a food delivery algorithm or a multinational insurance treaty, certain underlying principles persist. By understanding these, we can formulate a more universal risk management philosophy. Before presenting the formal framework and principles, we summarize in Table 2.1 some of the key risk factors and mitigation approaches across the diverse domains discussed. This serves as a bridge from analysis to action, distilling practical lessons.

Table 2.1 shows that many domains resort to similar tools, transparency, oversight, and better design to manage risk. This hints that a consolidated set of principles might underlie effective risk governance. In the next section, we articulate such a set of core principles as the crux of a unified framework and illustrate how they can be applied in practice with further global examples.

TOWARD A UNIFIED FRAMEWORK FOR SAFETY AND RISK

Drawing from the analysis above, we propose a unified framework grounded in a set of core principles that can guide risk management across diverse sectors. These principles are informed by both philosophical considerations (fairness, responsibility, harm avoidance) and practical evidence (what has worked in real-world case studies). The framework is meant to be holistic, considering technological, human, and institutional aspects, and actionable, offering concrete approaches rather than abstract slogans.

The five core principles of the framework are:

1. **Transparency and Truthfulness.** Ensure that accurate information flows to all stakeholders. Hidden risks and deceptive practices should be minimized by design.
2. **Alignment of Incentives.** Structure systems so that doing the "right" (safe/fair) thing is also in the actor's self-interest. Conversely, actors should bear an appropriate share of the consequences of the risks they create (internalizing externalities).
3. **Empowerment and Education.** Empower individuals with tools and knowledge to manage their own risk exposure according to their values, recognizing behavioral biases. This includes ethical design that supports user welfare.
4. **Responsibility and Accountability.** Clearly define who is responsible for what risk and hold those actors accountable for failures. This applies to both corporate responsibility (e.g., platform duty of care) and personal responsibility (e.g., user due diligence), balanced fairly.
5. **Adaptive Governance and Collaboration.** Use flexible, forward-looking regulation and cooperative approaches that can adapt to emerging risks and cross-domain interdependencies. Encourage dialogue between sectors and regularly update safeguards as conditions change.

These principles are mutually reinforcing. For example, transparency (Principle 1) often enables better personal decision-making (Principle 3) and eases regulatory oversight (Principle 5). Incentive alignment (Principle 2) often requires establishing clear accountability (Principle 4). Together, they create an environment where safety and risk are proactively managed rather than reactively lamented.

We detail each principle below, with illustrative instances of how they might be implemented in various fields, followed by Table 2.2, which summarizes the principles and provides cross-domain examples of their application.

Principle 1: Transparency and Truthfulness. This principle echoes the age-old philosophical commitment to truth, now applied to risk contexts. Practically, it means all critical information about risks, benefits, and system operations should be open or easily accessible. For example, in fintech lending, this means transparent credit scoring and loan terms. If machine learning algorithms are used to set interest rates or insurance premiums, there have been calls for "algorithmic transparency," explaining to consumers why they got a certain rate, which can also help detect bias. In food safety and supply chains, transparency is addressed by technologies like blockchain to trace food origin and handling conditions, reducing the risk of contamination by making every step visible and auditable. In fake reviews, transparency might involve platforms disclosing their verification process and success rates in removing fraud, thereby building user trust that the remaining reviews are likely genuine. Importantly, transparency extends to acknowledging uncertainties. Engineers are taught to communicate the limits of a model or the error bars on predictions; similarly, regulators should be frank about unknowns (e.g., openly discussing what is not

TABLE 2.2
Core Principles of the Unified Risk Framework with Cross-Domain Examples

Principle	Description (What It Means)	Example in Practice (Different Domains)
1. Transparency and Truthfulness	Provide clear, truthful information about risks and system operations to all stakeholders. Minimize information asymmetry and secrecy that could hide harms.	*Finance:* Banks must clearly disclose loan costs and risks (e.g., standardized "Schumer box" for credit card terms), no hidden fees or fine print traps. *Food Safety:* Restaurants display hygiene ratings publicly; food delivery apps show customers real-time info on food sourcing and delivery conditions (e.g., temperature) to assure safety.
2. Alignment of Incentives	Design incentives so that safe, fair outcomes align with the self-interest of actors. Those who create risk should bear costs; those who reduce risk should benefit.	*Insurance:* Usage-based car insurance (telematics) charges drivers based on driving behavior, safe driving lowers premiums, aligning cost with actual risk and rewarding safety. *Online Platforms:* E-commerce marketplaces make seller visibility contingent on low complaint rates, sellers who provide safe, quality products get more customers, incentivizing good practices (and unsafe products are algorithmically demoted).
3. Empowerment and Education	Empower individuals with tools and knowledge to understand and manage risks. Account for human behavior and biases in design, guiding users toward informed decisions without removing their agency.	*Gambling:* Casinos and betting apps implement voluntary self-exclusion and limit-setting features; players are educated on odds and signs of problem gambling. Some jurisdictions require pop-up warnings like "You have been playing for 1 hour" to prompt reflection. *Cybersecurity:* Regular public awareness campaigns teach people how to identify phishing and fraud (e.g., banks send out test scam emails to customers with "Did you spot this?" follow-ups to educate). Browsers empower users with warnings: "This site may be deceptive, go back?", giving a choice but nudging safety.

(*Continued*)

TABLE 2.2 (Continued)
Core Principles of the Unified Risk Framework with Cross-Domain Examples

4. Responsibility and Accountability	Clearly assign responsibility for risk prevention and mitigation. Hold parties accountable for negligence or harm through appropriate legal, financial, or reputational consequences. Ensure mechanisms for redress.	*Product Safety:* Manufacturers are legally liable for defects (product liability laws), meaning if a toaster catches fire due to design flaw, the company must compensate victims, incentivizing rigorous safety testing. *Online Content:* "Safe harbor" laws condition platform immunity on taking responsibility to remove illicit content once notified. New regulations (e.g., EU Digital Services Act) require large platforms to assess and mitigate systemic risks on their services, under threat of hefty fines, effectively making them accountable for algorithmic harms (like viral misinformation or dangerous challenges).
5. Adaptive Governance and Collaboration	Use flexible, evidence-based regulatory approaches that can adapt to new risks. Encourage collaboration across sectors and borders, sharing data and best practices. Continually monitor and update risk controls.	*Environmental Risk:* Climate risk task forces bring together scientists, policymakers, insurers, and community leaders to iteratively plan for climate resilience (e.g., updating flood maps and building codes every few years as data improves). *Financial Regulation:* Regulators employ "stress tests" for banks under hypothetical crisis scenarios and adjust capital requirements accordingly. They collaborate through international bodies (Basel Committee) to update rules in light of global insights, for example, after 2008 crisis, worldwide reforms (Basel III) were coordinated to prevent regulatory arbitrage. *Tech Ethics:* Cross, industry consortiums (e.g., Partnership on AI) involve multiple tech companies and NGOs working with governments to set adaptive guidelines on AI safety, as everyone learns from each other's mistakes and breakthroughs.

yet understood about long-term impacts of a new gambling product or a vaccine). A culture of transparency creates a feedback loop: stakeholders can only hold each other accountable (Principle 4) if they have the facts. One success story is the aviation industry's incident reporting systems: pilots and crew globally share incident reports in databases that are openly studied, enabling the whole industry to learn from near misses. Contrast that with healthcare, where historically there was a culture of hiding mistakes, which slowed safety improvements. Transparency can thus be lifesaving. Of course, there are limits; privacy and security concerns mean not everything can be open (e.g., one would not publish a bank's internal fraud detection algorithm or personal customer data). But even then, independent audits or oversight boards can provide surrogate transparency (they see the full picture and vouch for its integrity). In sum, when in doubt, more truthfulness and openness about risk tends to lead to better outcomes than secrecy or spin.

Principle 2: Alignment of Incentives. Echoing themes from utilitarian ethics and contract theory, this principle is about structuring "the game" so that individual rationality leads to collectively desirable results. A lot of harm comes when there are perverse incentives: e.g., if gig workers are paid per delivery, speed is incentivized at the expense of safety. A solution some have tried is paying by the hour plus a modest per delivery bonus, so that rushing yields diminishing returns. In finance, before 2008, mortgage brokers had incentives to approve loans and sell them off, regardless of quality; now regulations like "risk retention" require them to keep a stake, aligning incentives toward loan quality. Another example: in content moderation online, platforms originally had the incentive to ignore fake accounts boosting engagement; now, facing legal and reputational risk, their incentives are realigning toward removing such content. We should design incentive compatibility into systems from the start. For instance, if developing a new sharing economy app, one should ask: how can we make it profitable for us *and* safe for users? Perhaps by charging a fee that is tied to safety metrics or long-term customer satisfaction, rather than pure volume. At a policy level, governments use taxes and subsidies to align incentives with social good: taxing cigarettes (to account for health harms), or giving insurance discounts for safe drivers, etc. The principle also implies harnessing market forces for good. For example, insurance companies, in pursuing profit, naturally gather data on what reduces claims; this incentive can be leveraged by regulators to improve standards (as insurers will lobby for safer building codes if it reduces payouts). A unified framework encourages systematically analyzing each stakeholder's incentives and tweaking them. Philosophers might relate this to the idea of the "invisible hand" working properly only when the rules of the game are well designed to equate private and social costs/benefits. In practice, achieving perfect alignment is hard, but even partial moves (like performance-based regulation where companies get incentives for reducing accidents or defaults) can markedly reduce harm.

Principle 3: Empowerment and Education. People are the ultimate risk managers in their own lives. Thus, empowering individuals with both the tools and the understanding to make wise choices is vital. Education here isn't just formal, but any means by which people become more informed and capable regarding risks. For instance, financial literacy programs can help borrowers understand compound interest and avoid predatory loans; some countries have started integrating basic finance into school curricula. In the tech domain, digital literacy (how to spot a phishing email or a fake review) is being increasingly recognized as a life skill. Empowerment also involves giving users control: many online services now provide privacy settings or self-exclusion options (in gambling apps, users can set a cooling-off period). An empowered consumer base acts as a first line of defense; they are less likely to fall for scams or be reckless if they are well-informed and have tools to act on that knowledge. A case in point: countries like Sweden, which have robust driver education and public awareness around road safety, tend to have lower accident rates even before considering vehicle or infrastructure differences (education fosters a safety culture). However, empowerment must account for human limitations; simply handing out brochures is not enough. Innovative approaches include gamified learning (simulating consequences of risky actions in a safe environment) or peer-to-peer support networks (e.g., gambling addiction support groups that educate new members on pitfalls). Empowerment is also about equity: ensuring that all groups (including marginalized communities) have access to risk information and protective tools. Often, those most at risk (e.g., low-income borrowers, elderly internet users susceptible to scams) are the hardest to reach with traditional education, so creative outreach (community workshops, translated materials, etc.) is needed. In our framework, every intervention should ask: how does this help people help themselves? For example, if a regulator is considering mandating a warning label, will consumers understand it, or do they need a visual symbol? Should there be an interactive element (like a prompt saying "do you really want to bet more than your monthly limit?" that forces a moment of reflection)? By focusing on empowerment, we respect individual autonomy while still guiding toward safer behavior; it is a middle path between laissez-faire and heavy paternalism.

Principle 4: Responsibility and Accountability. This principle embodies the ethical notion that with power or benefit comes responsibility. It insists on clearly delineating who is accountable for preventing, mitigating, or compensating for harms. In complex systems, there is often a dilution of responsibility ("many hands" problem). A gig platform might say "we just connect customers and couriers, we are not responsible for accidents"; a social media company might claim "we are not publishers, so we are not responsible for misinformation." The framework, however, encourages assigning accountability in a way that incentives to control risk are maximized. If a platform has significant control over the transactions and profits from them, it stands to reason they carry responsibility for safety issues on that platform. Regulators worldwide are moving in this direction; for example, several countries have enacted "duty of care" laws requiring online platforms to actively manage harmful content and not just react when prompted. Accountability also means learning from failures: when disasters or scandals happen, holding inquiries and public reports to identify mistakes and assign liability is crucial. This not only provides justice but also deters negligence. Legal liability (civil or criminal) is a blunt but effective tool to enforce accountability, e.g., heavy fines on financial institutions for mis-selling risky products or jail time for executives who commit fraud (which is a risk to market integrity). In engineering, a culture of accountability is fostered through professional ethics codes: e.g., civil engineers can lose licenses for safety lapses. Similarly, financial advisors can lose certifications for violating fiduciary duty. Accountability can also be shared: public-private partnerships might share responsibility for infrastructure risks (government takes some because it sets regulations; companies take some because they do execution). Importantly, accountability is not only top-down; consumer accountability exists too, e.g., someone who knowingly lies on a loan application bears responsibility for resulting issues. Our framework supports fair apportionment: do not punish agents for things out of their control but do hold them accountable for the risks they create or could reasonably prevent.

Principle 5: Adaptive Governance and Collaboration. In a fast-changing risk landscape, no single entity has all the answers. Governance needs to be flexible and inclusive of various expertise. Adaptive governance takes cues from adaptive management in ecology; policies are treated as hypotheses to be tested and tweaked. For instance, a new regulation on crypto exchanges could be introduced with a sunset clause and a plan to review outcomes after two years, adjusting based on evidence. This prevents ossification of bad rules or loopholes. Collaboration is also paramount: many of the problems we discussed cross traditional boundaries (public vs private sector, national borders, different industries). Thus, solutions often require multi-stakeholder efforts. A contemporary example is in cybersecurity: governments, tech companies, financial institutions, and even civil society groups collaborate in information-sharing hubs to rapidly respond to cyber threats, acknowledging that cyber risk is a shared burden. Similarly, in addressing, say, fake reviews, enforcement might involve global cooperation (the FTC rule will be most effective if e-commerce sites worldwide adopt similar stances; otherwise, fake review sellers will just move to other markets). International standards bodies (like ISO for safety management or the Financial Stability Board for banking) are forms of collaborative governance that help harmonize approaches across borders, which is increasingly necessary. The principle also endorses involving the public in governance (e.g., participatory approaches, citizen juries for local risk decisions) to ensure policies have legitimacy and tap local knowledge. An example of adaptiveness is how some regulators now employ "RegTech," using AI tools themselves to monitor markets in real time (rather than static audits). This allows quicker reaction to, say, a spike in default rates or a new gambling product trend. Essentially, Principle 5 is about learning and adjusting, a humility in governance that we don't have a perfect static solution, so we remain vigilant and ready to change course if needed. It also emphasizes breaking silos: engineers talking to sociologists when implementing a safety measure or financial analysts consulting psychologists when crafting disclosure forms (to ensure they truly inform). By fostering a culture of continuous improvement and teamwork across disciplines, we approach the ideal of harmony, dynamic equilibrium rather than zero risk (which is impossible). We accept some risk but manage it in a way that society can absorb and bounce back (resilience).

These five principles form a cohesive framework. To illustrate them concretely, Table 2.2 presents each principle with an example of how it can be applied across a couple of different domains (showing its versatility).

By embedding these principles, organizations and regulators create a robust safety culture that transcends any one domain. This unified framework does not imply a one-size-fits-all solution; rather, it provides a philosophical and practical lens through which to approach risk in any context. An engineer can ask, "Am I being transparent about the uncertainties in my design? Have I aligned incentives for users to use this safely? Have I accounted for human error?" A policymaker can ask, "Who is accountable if this goes wrong? Are we giving people the knowledge to navigate this safely? How will this policy hold up if conditions change?" These questions, rooted in the five principles, lead to more resilient and ethical outcomes.

IMPLICATIONS FOR ENGINEERS, REGULATORS, AND PRACTITIONERS

Having outlined the framework, we turn to its real-world application. What do the above principles mean for those who design systems (engineers and product developers), those who set the rules (regulators and policymakers), and those on the ground running operations or services (practitioners, including business managers and safety professionals)? In this section, we discuss the implications for each group, recognizing that in practice their roles often overlap, and they must work together.

- Engineers and Designers: Engineers are often the architects of risk controls (or, if unaware, the creators of risk externalities). The framework urges engineers to adopt a safety-by-design mentality. This means proactively integrating safety features and fail-safes rather than seeing them as afterthoughts. For example, software engineers building a fintech app should implement safeguards for users making high-stakes transactions (like "Are you sure? This is 50% of your account balance"), reflecting Principles 3 and 1. System designers should also use human-centric design, acknowledging that users might misuse or not understand a feature, so make it hard to accidentally do something dangerous and easy to do the safe thing. Engineers must also embrace interdisciplinary input: collaborating with domain experts (finance, medicine, ethics) to foresee consequences beyond the technical. Importantly, transparency (Principle 1) for engineers means making systems interpretable and auditable. In AI development, for instance, the push is for "explainable AI" where decisions can be understood by humans. Engineers should document assumptions and limitations in their models. The framework also implies engineers consider ethical implications: there is increasing expectation (and in some cases legal requirement) that engineers follow codes like medical ethics, e.g., the IEEE has an ethical standard for autonomous systems. Overall, engineers should view themselves as custodians of public trust: when they build a bridge, a payment system, or an app algorithm, they hold lives and livelihoods in their hands by extension. The chapter's insights into various failures (from Boeing's 737 Max crashes to algorithmic biases causing discrimination) underscore that technical brilliance must be coupled with ethical and risk awareness.

- Regulators and Policymakers: For regulators, the unified framework provides a checklist for comprehensive oversight. Regulators should strive to create a regulatory ecosystem that embodies the five principles. Concretely, this might mean updating laws to mandate transparency (e.g., requiring platforms to report how they moderate content and with what error rates), setting up mechanisms that align business incentives with public interest (like performance-based regulation or liability schemes), and investing in public education (perhaps using fines from violators to fund consumer awareness programs). Policymakers must also improve their agility: traditional rulemaking can be glacial, but modern risks evolve fast (witness how crypto innovations outpaced financial regulations). Approaches like "regulatory sandboxes" and periodic sunset reviews of rules can help keep regulations relevant. Collaboration is vital: regulators should communicate with each other (e.g., a fintech regulator working with a data protection authority on overlapping issues) and internationally (since risks like pandemics or financial crises are global). A significant implication is the need for capacity building; regulators need expertise in data science, human factors, etc., not just legal knowledge. Many agencies are hiring data analysts, psychologists, and ethicists to better understand the industries they regulate. The framework also nudges policymakers to think holistically about risk trade-offs: a policy addressing one risk might create another, so impact assessments should be broad. For instance, heavily restricting certain loans might push vulnerable borrowers to even riskier informal lenders; thus, a balanced approach like offering safer alternatives is wiser. In philosophy of governance, this aligns with the idea of paternalism vs autonomy; regulators must carefully decide when to protect people from themselves (like mandating motorcycle helmets) versus when to let choice prevail but provide support (like making helmets optional but providing education and incentives). The evidence from different domains can guide these choices (helmet mandates, for example, clearly saved lives in traffic safety, whereas prohibition in gambling historically led to black markets, suggesting regulated availability with protections is better). Finally, regulators should consider metrics of success beyond traditional economic measures: for example, measure public trust, measure the reduction in harm incidents, not just industry growth. A safety-oriented culture in regulation means success is when things *don't* go wrong, which requires a proactive mindset.
- Industry Practitioners and Managers: This group includes business leaders, risk officers, safety managers, anyone responsible for operationalizing risk management. The unified framework suggests that integrating risk considerations into core strategy is not only ethically sound but often profitable long-term. For instance, a bank that builds a reputation for fair dealing and not exploiting customers may sacrifice some short-term fee income but gains loyalty and avoids legal troubles, improving long-term value. Practitioners should implement robust risk governance within organizations: this means clear policies, internal audits, whistleblower channels,

and a culture where safety or ethics concerns can be raised without fear. As seen in cases like the Wells Fargo fake accounts scandal (where employees created fake accounts to meet incentives), a toxic incentive culture can cause massive harm; companies must be mindful to align internal incentives with desired outcomes (Principle 2 applied internally). Managers in sectors like manufacturing or transport should also seek cross-industry learning, e.g., the healthcare industry borrowing safety practices from aviation (checklists, blame-free error reporting) has improved outcomes. The chapter's interdisciplinary approach suggests that innovation in safety can come from anywhere: a casino might learn from e-commerce fraud detection to spot problem gambling patterns or a food delivery company might emulate airline fatigue management for its drivers. Practitioners should also engage with regulators constructively: those who participate in developing standards or pilot new compliance tech can shape regulations in mutually beneficial ways. Ethical leadership is another implication: leaders need to set the tone that "how" profits are made matters, not just how much. In practical terms, that could mean a CEO deciding to pull a profitable product off the market upon evidence it is unsafe (as Johnson & Johnson did in the Tylenol poisoning crisis, which ironically increased trust in the brand). For small businesses or startups, the framework is a reminder to bake in good practices early (it's harder to retrofit ethics and safety). Many startups now include a risk officer or at least consult compliance experts from the get-go, especially in fintech and biotech. From a global perspective, practitioners in different regions should share knowledge, e.g., a payment app in Kenya (M-Pesa) taught the world how mobile money could be done safely among low-literacy users via community agents, an insight now used elsewhere. The harmony concept implies that doing good risk management can be a competitive advantage; it is part of the value proposition to customers (they can trust you), to investors (reduced legal and reputational risk), and to employees (who feel safer and prouder of their work).

CONCLUSION

In this chapter, we have journeyed through a panorama of modern risks, from the gig economy's bicycle-laden food couriers to the intangible algorithms trading cryptocurrency at lightning speed, and we have sought to find the common philosophical threads and practical strategies that can guide us toward a safer, more harmonious coexistence with risk. The title *"From Harm to Harmony"* is an aspirational arc: it recognizes that risk can never be eliminated (to live is to face uncertainty), but it posits that through knowledge, ethics, and prudent governance, we can transform potential harms into manageable, even beneficial, aspects of human progress.

We began by examining a set of seemingly disparate domains, each highlighting different facets of risk: physical safety (food delivery riders braving traffic), financial security (borrowers juggling car loans or gamblers chasing losses), technological trust (crypto investors and online consumers navigating new frontiers), and informational integrity (fake reviews polluting our collective decision-making channels). This examination underscored that while contexts differ, the patterns of risk, and often the sources of failure, are strikingly similar. The lack of transparency, misaligned incentives, neglect of human factors, fragmented oversight, and unaccountable power appear again as the underlying causes when risk spirals into harm.

From this insight, we distilled a unified framework embodied in five core principles: Transparency, Alignment, Empowerment, Responsibility, and Adaptability. These principles are not radically new; indeed, they echo long-standing ideals in philosophy and governance (from Aristotle's call for virtue in conduct to Enlightenment ideas of accountability to modern systems theory). What is new is applying them in a synthesized, interdisciplinary manner to the emergent challenges of the twenty-first century. Whether the challenge is how to make a loyalty program fair or how to regulate artificial intelligence, these principles provide a robust checklist and value set. They remind us to ask: Are we being honest and open? Are we structuring this so that doing the right thing is incentivized? Are we equipping people to make good choices? Have we clearly assigned who is responsible for what? And are we ready to learn and adjust as we go?

The global case studies and examples scattered through the chapter give life to these principles. We saw how regulators in the United States acted against fake reviews to protect consumers and honest businesses, a move toward transparency and accountability in the digital marketplace. We noted how an international commission on gambling called for treating gambling as a public health matter, exemplifying adaptive governance and cross-sector collaboration (health, finance, law enforcement working together). We discussed how technology, like blockchain, is being leveraged for food safety, a case of transparency through innovation. We reflected on the lessons of financial crises, which taught hard lessons about incentive misalignments and lack of oversight, spurring reforms that align banking practices more with societal stability. We also highlighted positive strides: for instance, the high insurance penetration in some countries that helped absorb disaster losses is a success story of risk pooling and proactive policy. These examples collectively show that progress is possible. Many industries today are far safer and more consumer-friendly than a few decades ago because of conscious efforts to implement such principles.

Yet, we must be candid that challenges remain and new ones will emerge. The rise of AI, biotech, and other technologies will pose novel ethical and safety dilemmas. Climate change is introducing systemic risks that our current institutions are only beginning to grasp. Geopolitical shifts can undo cooperative safety nets (as seen when international coordination falters). Moreover, the balance between freedom and safety will continuously be debated; one person's acceptable risk is another's intolerable harm. Philosophers will still be needed to navigate these gray areas: how much paternalism is justified to prevent self-harm? How do we weigh present costs against future risks (a classic dilemma in climate policy and beyond)? Ensuring that equity is at the heart of risk management is another ongoing task; historically, vulnerable communities have borne disproportionate harms (whether poorer countries facing the brunt of climate disasters or marginalized groups targeted by predatory financial schemes). A harmonious approach to risk must strive to correct these imbalances, making safety not a privilege but a shared standard.

For engineers, regulators, and practitioners, the work is never "finished." A safety system is not a monument, it is a garden; it requires tending, vigilance, and adaptation. This chapter and the book advocate for a mindset of continuous improvement anchored in sound principles. It encourages an ongoing dialogue between theory and practice: philosophical perspectives sharpen our sense of what ultimately matters (human dignity, justice, flourishing), while hands-on case studies keep us grounded in what is effective or not in the real world.

In concluding, we circle back to the metaphor of harmony. In a harmonious orchestra, many different instruments (domains, stakeholders) play in sync under a guiding melody (principles and goals). Dissonance (risk) is not eliminated; in fact, a measure of risk is the spice of innovation and human endeavor, but it is resolved within a broader concord. Achieving this is not trivial; it requires leadership with vision, public engagement, ethical courage, and technical expertise. However, as we have argued, it is achievable. We see glimpses of it wherever thoughtful risk governance has been applied. A future where food delivery riders zip through cities safely, where financial products empower rather than exploit, where digital spaces are trustworthy, and where innovation thrives without routinely stepping over the precipice of disaster—that future is the one we must intentionally craft.

REFERENCES

1. Jarvis, D.S., 2007. Risk, globalisation and the state: A critical appraisal of Ulrich Beck and the world risk society thesis. *Global Society*, 21(1), pp.23–46.
2. Hansson, S.O., 2004. Philosophical perspectives on risk. *Techné: Research in Philosophy and Technology*, 8(1), pp.10–35.
3. Yazdi, M., 2025. From harm to harmony: What philosophers teach us about safety. In Yazdi, M. (eds). *Safety-Centric Operations Research: Innovations and Integrative Approaches*, Studies in Systems, Decision and Control. 232, Springer, Cham. https://doi.org/10.1007/978-3-031-82934-5_1.
4. Stirling, A., 2007. Risk, precaution and science: Towards a more constructive policy debate: Talking point on the precautionary principle. *EMBO Reports*, 8(4), pp.309–315.
5. Rusli, R., Amri, F.H.S., Yahya, N.A., Oviedo-Trespalacios, O., Rahim, S.A.S.M., Ismail, A.A. and Salim, P.I.S., 2025. Rush hours: The risky behaviours of food delivery and non-food delivery motorcycle riders. *Journal of Road Safety*, 36(3), pp.27–35.
6. Bakker, T.J., DeLuca, S., English, E.A., Fogel, J.S., Hendren, N. and Herbst, D., 2025. Credit access in the united states (No. w34053). National Bureau of Economic Research.
7. Seide, C.K., 2012. Consumer financial protection post Dodd-Frank: Solutions to protect consumers against wrongful foreclosure practices and predatory subprime auto lending. *UPR Business Law Journal*, 3, p.219.
8. Yendork, J.S., Wardle, H., Degenhardt, L., Marionneau, V. and Reith, G., 2024. The Lancet Public Health Commission on gambling.
9. Kita, S.M., 2025. Disaster insurance for climate loss and damage: Assessing challenges and opportunities to adoption in Malawi. *Risk, Hazards & Crisis in Public Policy*, 16(1), p.e70002.

10. Agrawal, R., Khanna, A. and Hamdare, S., 2025. Analyzing and rewarding credit card spending habits in India: A machine learning approach. *International Journal of Computational Intelligence Systems*, 18(1), p.165.
11. Kucinskas, J.L., 2025. *The Loyalty Trap: Conflicting Loyalties of Civil Servants under Increasing Autocracy*. New York, Columbia University Press.
12. McLauchlin, T., 2018. The loyalty trap: Regime ethnic exclusion, commitment problems, and civil war duration in Syria and beyond. *Security Studies*, 27(2), pp.296–317.
13. Arner, D.W., Zetzsche, D.A., Buckley, R.P. and Kirkwood, J.M., 2023. The financialization of crypto: Lessons from FTX and the crypto winter of 2022–2023. University of Hong Kong Faculty of Law Research Paper, (2023/19), pp.23–31.
14. Johnson, R. and Cureton, A., 2004. Kant's moral philosophy.
15. Rizwan, M.S., Ahmad, G. and Ashraf, D., 2020. Systemic risk: The impact of COVID-19. *Finance Research Letters*, 36, p.101682.

3 Technological Ethos and Safety Innovations

Philosophical Reflections on the Ethical Integration of Emerging Technologies in Safety Practice

INTRODUCTION: WHY "TECHNOLOGICAL ETHOS" NOW MATTERS FOR SAFETY

Safety practice has always been shaped by technology. Guarding, interlocks, sensors, and control systems changed what "good safety" looked like long before AI. The difference today is speed, reach, and *decision delegation* [1]. AI-enabled systems increasingly recommend actions, prioritize risks, and classify events, sometimes faster than humans can verify, and sometimes with limited transparency. At the same time, the strategic narrative around "Industry 5.0" explicitly re-centers human wellbeing, sustainability, and resilience, pushing against a purely efficiency-driven logic of the prior industrial wave.

Technological ethos is the underlying moral "direction" of technology in practice: what the organization treats as legitimate goals (e.g., productivity vs dignity), what it counts as valid evidence, who is empowered to decide, and which risks are ignored because they are inconvenient. In safety, technological ethos shows up in questions such as:

- Is technology used to *support* workers' judgment or to *replace* it?
- Are workers treated as partners in risk control, or as data sources to be monitored?
- Do algorithms increase organizational learning, or create new blame pathways by hiding decision logic?

Safety thinking has also evolved toward human-centric approaches. In the Safety 5.0 discussion, tensions are highlighted between AI-steered processes and human control, and the changing profile of threats, from primarily physical harm toward stress and psychosocial risk, especially in technology-mediated work [2].

DOI: 10.1201/9781003634720-3

This chapter therefore addresses two linked tasks:

1. examining how AI and automation reshape traditional safety paradigms, and
2. proposing a straightforward framework for ethically integrating new technologies into safety practice.

PHILOSOPHICAL LENSES THAT CLARIFY ETHICAL RISK IN SAFETY INNOVATION

Ethics is often treated as a compliance add-on ("privacy policy," "consent form"); however, philosophical approaches help safety leaders see *structural* problems, especially those produced by power asymmetry and opaque decision-making.

Deontology: Dignity, Rights, and "People As Ends"

A deontological lens asks whether a system respects persons, not merely whether it improves metrics. A core Kantian idea is to avoid treating people as *mere means* to organizational ends (e.g., extracting behavioral data without meaningful choice or forcing algorithmic schedules that ignore human limits) [3]. In safety terms, this reinforces worker dignity, participation, and non-coercive oversight of monitoring technologies.

Consequentialism: Benefits, Harms, and Distribution

A consequential lens is natural for safety (risk reduction, harm prevention). However, it becomes ethically incomplete if it ignores **distribution**, who receives safety benefits, who bears surveillance burdens, and whether risks shift from visible injuries to hidden psychosocial harms.

Virtue/Practical Wisdom: "Deskilling" and Moral Outsourcing

Safety competence includes judgment under uncertainty. Automation can unintentionally weaken that competence when people over-trust decision aids, stop cross-checking, or lose situational awareness. Research on complacency and automation bias shows these effects are persistent and not easily eliminated by simple training, especially under workload and imperfect automation [4]. This matters because safety is not just rules; it is cultivated practical wisdom.

Socio-Technical Ethics: Technology Is Never "Neutral" in Practice

A socio-technical lens treats safety as emerging from interactions among humans, procedures, incentives, and technical systems. Under this lens, "ethical integration" means designing controls that address incentives, governance, and accountability, not only model accuracy.

HOW EMERGING TECHNOLOGIES RESHAPE SAFETY PARADIGMS

AI/ML: From Lagging Indicators to Predictive Control, And New Opacity

AI can detect patterns in near misses, maintenance logs, and sensor streams, enabling earlier intervention. Yet predictive systems can also become **epistemic authorities** ("the model says it's safe"), pushing teams toward complacency. When a system's reasoning is not explainable, it can weaken learning, because organizations cannot easily ask: *why did this happen?* and *what should change?*

A useful anchor is that trustworthy AI must be lawful, ethical, and robust, and it should meet requirements including human oversight, technical robustness and safety, privacy/data governance, transparency, and fairness [5]. Those requirements map directly to safety system needs (risk control, competence, change management, and verification).

Robotics and Automation: Fewer Injuries, but New Psychosocial and Systemic Risks

Automation can remove workers from high-energy hazards. Evidence using establishment-level injury data suggests that increasing exposure to industrial robots is associated with lower injury rates; for example, one study reports that a one standard deviation increase in robot exposure reduces annual workplace injury rates by about **1.2 cases per 100 workers** [6].

However, the same research stream warns of second-order harms: regions more exposed to robot penetration can show increases in drug-/alcohol-related deaths and mental health problems. This illustrates a key Safety 5.0 lesson: innovations can reduce *physical* harm while increasing *psychosocial* harm if the technological ethos prioritizes productivity without sufficient human-centered safeguards.

Collaborative robots (cobots) also introduce shared workspaces and contact hazards. Safety guidance around collaborative operation emphasizes risk assessment, force/pressure limits, and engineered safeguards (often discussed alongside ISO/TS 15066 concepts) [7].

Wearables and Lot: Strong Leading Indicators, High Privacy Stakes

Wearables can support ergonomic risk management, fatigue detection, and early warnings (e.g., unsafe posture, overexertion). Real-world case evidence is increasingly used to justify adoption. For example, a logistics/records management case study reported:

- **64% reduction** in OSHA recordable injury rates (year-over-year);
- **58% reduction** in workers' compensation claim costs;
- **5:1** ratio of dollars saved to dollars invested after integrating wearable tech with safety training and analytics [8].

These results are attractive for safety leaders; however, wearables intensify ethical duties around consent, proportionality (collect only what you need), purpose limitation, and protection from misuse (e.g., disciplinary surveillance rather than harm prevention).

Digital Twins: Safety Simulation at Scale and Ethical Questions About Ownership and Trust

Digital twins can simulate equipment degradation, process deviations, and even human-system interactions. Yet public-sector technology assessments note challenges including privacy and ethical issues (security, bias, data ownership), technical barriers to scaling, and potentially high costs that can create inequities [9].

The ethical point is simple: when models represent people, workplaces, or performance, they can become contested sources of truth. Governance must define what the twin is allowed to decide, how it is validated, and who can challenge it.

Generative AI: Fast Procedure Drafting, High Hallucination Risk

Generative AI can draft SOPs, toolbox talks, and risk assessments quickly. However, safety is high-stakes: plausible-but-wrong outputs can introduce latent hazards. Ethical integration here means *treating generative AI as a drafting aid*, not an authority, requiring human verification, traceable sources, and strong change control.

ETHICAL RISK CATEGORIES THAT MATTER MOST IN SAFETY TECH

Ethical issues become safety issues when they change behavior, weaken learning, or shift control away from accountable humans. The following categories should be explicitly assessed in any safety technology project.

Human Agency and Oversight

If a system can meaningfully affect job allocation, performance evaluation, or safety decisions, "human-in-command" mechanisms are required. Trustworthy AI frameworks explicitly emphasize human agency and oversight as a core requirement, including human-in-the-loop/on-the-loop approaches.

Transparency, Explainability, and Contestability

Opaque models weaken incident learning and fair decision-making. Transparency does not require revealing proprietary code; it requires meaningful explanations to affected stakeholders (workers, HSRs, supervisors) about system capability and limits, and a practical way to challenge outputs.

Privacy, Surveillance Creep, and Data Ownership

Digital twin assessments highlight privacy, data ownership, and trust concerns [9]. Wearables raise parallel issues. The ethical principle: safety data must not become a backdoor HR surveillance tool unless explicitly justified, proportionate, and negotiated.

Bias and Unfairness

Bias can systematically misclassify risk for particular groups (e.g., smaller body sizes, different gait patterns, culturally diverse communication styles), or over-target certain workers for "unsafe behavior" flags, creating injustice and undermining safety culture.

Automation Bias and Complacency

Evidence on automation bias indicates that users may commit omission/commission errors when decision aids are imperfect, and these effects can persist even among experts and are not fully prevented by training alone [4]. That means ethical integration must include *system design* (e.g., uncertainty display, forced cross-checks), not just "user training."

Cyber-Physical Risk Pathways

As safety tools become connected, cybersecurity failures can manifest as physical harm (e.g., manipulated sensor readings, disabled alerts). Safety tech governance must therefore include security-by-design and incident response integration.

PRACTICAL MAPPING: TECHNOLOGIES, BENEFITS, ETHICAL RISKS, AND CONTROLS

Table 3.1 provides a straightforward "risk-control" lens that combines safety and ethics. Use it during concept selection and procurement.

A FRAMEWORK FOR ETHICALLY INTEGRATING SAFETY TECHNOLOGIES: ETHOS-SAFE

This chapter proposes **ETHOS-SAFE**, a lifecycle framework designed to be usable by safety teams without requiring deep AI expertise. It aligns with widely used "trustworthy AI" principles and with risk management logic.

ETHOS-SAFE: Phases and Outputs

Phase 1: Purpose and Values (Ethos Check)

- Define what "success" means beyond injury metrics: dignity, fairness, worker participation, learning capacity.
- State non-negotiables: e.g., "No disciplinary use of wearable data," "Human-in-command for high-impact decisions."

TABLE 3.1
Technology–Ethics–Safety Mapping (Use for Design Reviews and Procurement)

Technology	Typical Safety Value	Primary Ethical/safety Risks	Minimum Controls (Non-negotiable)
AI risk scoring / prediction	Earlier detection of emerging risk; prioritization of interventions	Opaque decisions; bias; over-reliance; accountability gaps	Human oversight; explainability; bias testing; uncertainty communication; audit trails; change control
Robotics / cobots	Reduce exposure to hazardous tasks; consistency	Contact hazards; complacency; job stress; psychosocial harm	Task-based risk assessment; engineered guarding/limits; training; workload redesign; monitoring of psychosocial indicators
Wearables / IoT	Leading indicators (ergonomics/ fatigue); coaching	Privacy intrusion; disciplinary misuse; consent coercion	Purpose limitation; data minimization; governance with worker reps; secure storage; clear "no punitive use" rules
Digital twins	Simulation, scenario testing, proactive maintenance	Data ownership disputes; model bias; high cost inequity; false confidence	Validation plan; governance for decision rights; transparency about assumptions; access controls
GenAI for SOPs/JSAs	Fast drafting; consistency	Hallucinations; false authority; source ambiguity	Mandatory human verification; citeable sources; version control; restricted use cases

Phase 2: Context and Stakeholder Mapping

- Identify affected groups (including labor hire, contractors, vulnerable cohorts).
- Identify power asymmetries (who can say no, who can challenge outputs).

Phase 3: Dual Risk Assessment (Safety + Ethics)

- Traditional hazard identification plus ethical impact analysis: privacy, fairness, autonomy, contestability.
- Include "second-order harm" (stress, workload intensification), consistent with Safety 5.0 concerns about stress replacing physical threats in some contexts [2].

Phase 4: Design and Control Selection

Controls should follow hierarchy logic (eliminate, substitute, engineer, admin, PPE) **plus** data governance controls (minimize, secure, restrict access, log use).

Phase 5: Verification, Validation, and Assurance

- Technical: accuracy, robustness, drift monitoring, adversarial testing (where relevant).
- Social: usability, comprehension, trust calibration, fairness.
- Governance: clear accountability for overrides and failures.

Phase 6: Deployment and Change Management

- Competency development: operators, supervisors, safety team.
- Communication: capability/limitations and what data are (and aren't) used for.

Phase 7: Monitoring and Learning

- Continuous review: leading indicators, near misses, user feedback, psychosocial harm signals.
- Model drift and policy drift checks.

Phase 8: Decommissioning and Accountability

- Exit plan: data retention limits, safe shutdown, lessons learned, and preventing "shadow use" post-project.

ETHOS-SAFE Aligned to Established Governance

- EU trustworthy AI requirements offer concrete ethical categories: human oversight, robustness, privacy/data governance, transparency, fairness, etc. [5].
- NIST AI RMF provides a risk management orientation for AI trustworthiness (voluntary; however, widely adopted), supporting structured governance and lifecycle control [10].

DATA YOU CAN USE TO DRAW FIGURES (EVIDENCE AND BENCHMARKING)

Safety technology proposals often fail because teams cannot translate "ethics" into measurable outcomes. The goal here is not to reduce ethics to numbers; however, to ensure governance has *signals.*

Figure-Ready Dataset A: Selected Quantified Outcomes from Published Sources

The dataset below lets you draw a bar chart (percent reductions) and/or a mixed chart (absolute injury-rate change vs percent change). Use it as evidence in your chapter (Figure 3.3).

TABLE 3.2
ETHOS-SAFE Checklist Outputs (What to Produce, not just "Discuss")

Phase	Minimum Output	Owner	Review Trigger
1 Purpose/ethos	Tech charter + non-negotiables	Exec + HSE lead	Before vendor selection
3 Dual risk assessment	Safety risk register + Ethical impact register	HSE + worker reps	Before pilot
5 Validation	Assurance report (technical + social + governance)	Project + QA	Before go-live
7 Monitoring	KPI dashboard + drift/incident log	Ops + HSE	Monthly + post-incident
8 Decommission	Closure report + data disposal confirmation	Ops + IT + Legal	Project end

TABLE 3.3
Figure-Ready Dataset (Directly Reported Effect Sizes and Outcomes)

Technology/Intervention	Metric (Reported)	Quantified Outcome	Notes for Plotting
Industrial robots exposure	Work-related annual injury rate	**–1.2 cases per 100 workers** (per 1 SD increase in robot exposure)	Absolute change; plot as "injuries prevented per 100 workers"
Wearables + analytics + training	OSHA recordable injury rate	**–64%** year-over-year	Plot as percent reduction
Wearables + program	Workers' comp claim cost	**–58%**	Plot as percent reduction
Wearables program ROI	Dollars saved : invested	**5 : 1**	Plot as ROI ratio
National WHS benchmark (Australia)	Worker fatalities (count)	**200 (2023)**; **188 (2024)**	Plot counts over time (context baseline)

Robot and wearable outcomes are reported in the cited studies/case materials; Australian fatality counts are reported in national statistics summaries.

Figure-Ready Dataset B: A Simple Trend Table for Context

Table 2.4 provides a concise, figure-ready dataset that can be directly used to generate a line chart illustrating recent trends in traumatic injury fatalities among Australian workers. Presenting both the annual fatality count and the corresponding fatality rate per 100,000 workers helps establish the broader context for why ethically grounded,

TABLE 3.4
Australia Traumatic Injury Fatalities Trend Points

Year	Worker Fatalities (Count)	Fatality Rate (per 100,000 Workers)
2022	195	1:4
2023	200	1.4
2024	188	1.3

evidence-based innovation in work health and safety is essential. Even minor year-to-year fluctuations in fatality numbers highlight the persistent challenge of preventing serious harm and the ongoing need for solutions that are effective, responsible, and aligned with social expectations. The dataset focuses on the three most recent years for which national summaries are available. These values are drawn from Safe Work Australia's publicly released annual statistics on worker fatalities, which compile accepted workers' compensation claims and other verified fatality reports. Although the underlying causes of year-to-year changes vary, presenting the data in a simple trend form allows readers to quickly grasp the scale of the issue and understand the rationale for prioritizing safe and ethical innovation in high-risk industries.

As an example based on data from (https://data.safeworkaustralia.gov.au), Figure 3.1 presents the long-term trend in Australia's worker traumatic injury fatalities, showing both the annual number of fatalities and the fatality rate per 100,000 workers from 2003 to 2024. The gray bars illustrate the number of workers fatally injured each year, while the red line indicates the fatality rate adjusted for workforce size. The figure highlights a clear decline from the mid-2000s through the late-2010s, followed by a period of plateau where year-to-year improvements have slowed. This stabilization in recent years reinforces the importance of developing ethical, effective safety innovations that can help break through this plateau and continue reducing harm.

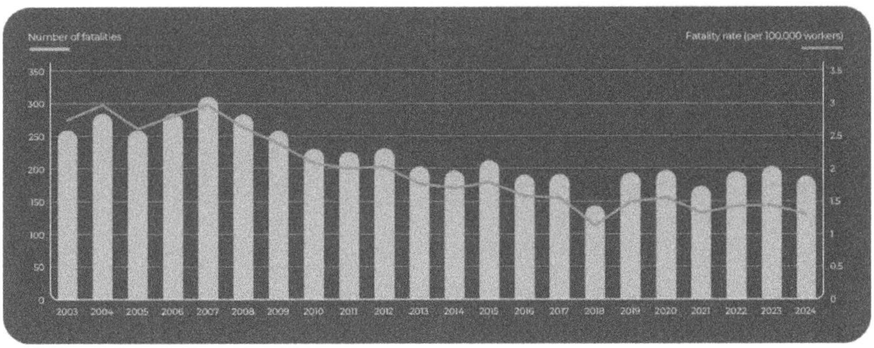

FIGURE 3.1 Trends in worker traumatic injury fatalities, Australia, 2003–2024.

As it can be seen from Figure 3.2, in 2024, worker fatalities in Australia were most commonly associated with vehicle incidents, which accounted for 66% (124) of all deaths. Vehicle incidents include situations where a worker, as an occupant of a vehicle such as a car, truck, aircraft, boat, tractor, quad bike, or other machinery, was killed following a collision with another vehicle or a stationary object. "Being hit by moving objects" also remained a significant mechanism of fatal injury, encompassing cases where workers on foot were struck by vehicles as well as injuries caused by moving equipment or other objects. Although the proportion of vehicle-related fatalities decreased slightly from 68% (139) in 2023, vehicles continue to dominate fatality mechanisms, with half of these incidents (50% or 62 deaths) involving at least one truck. Single-vehicle events made up the majority of vehicle-involved fatalities (72% or 89 deaths), indicating that loss of control, rollover, or other isolated incidents are more common than multi-vehicle collisions. Younger workers were disproportionately represented, with those under 25 experiencing the highest share of fatalities involving at least one vehicle (83%), while workers aged 25–34 had the lowest proportion (56%). Fatality rates also differed markedly by occupation: machinery operators and drivers recorded the highest fatality rate at 6.7 deaths per 100,000 workers, more than five times the national rate of 1.3, and accounted for 32% of all worker fatalities. Laborers recorded a rate of 4.0 per 100,000, while managers experienced a lower, however, still notable rate of 1.5 per 100,000 workers. Together, these patterns highlight the continuing role of vehicle-related risks and high-exposure occupations in Australia's fatal injury burden, underscoring the need for targeted, evidence-based safety interventions.

Figure 3.2 shows the distribution of outcomes for workers who experienced a work-related injury in 2024, distinguishing between those who received workers' compensation and those who did not. Overall, 30.5% of injured workers received compensation, while the remaining 69.5% did not. Among workers who did receive compensation, most required time away from work: 12.1% took between part of

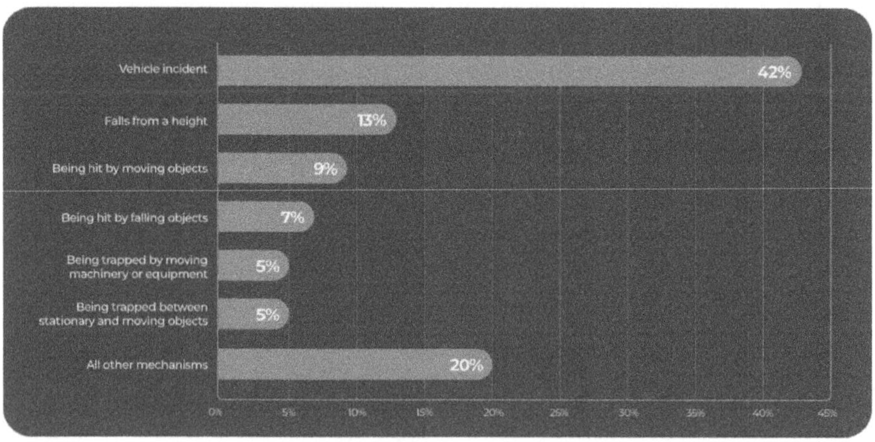

FIGURE 3.2 Worker fatalities by mechanism of incident, Australia 2024.

a day and 10 days off, and 14.1% took more than 10 days off or did not return to work. Only a small share, 4.3%, reported no time away from work despite receiving compensation, suggesting that injuries warranting compensation are generally more severe or require recovery time. For injured workers who did not receive workers' compensation, a different pattern is visible. A larger proportion (29.6%) took no time off work, and 29.4% took between part of a day and 10 days off, indicating these injuries were typically less severe or less disruptive. A smaller group (10.4%) required more than ten days away from work or did not return, despite not receiving compensation. Together, these results highlight that compensation status is closely linked to injury severity and recovery time, and that a substantial share of workers continue to manage injuries without accessing formal compensation pathways.

Figure 3.4 shows that the work-related injury and illness patterns differ notably across demographic groups, particularly by sex. In 2023–24, men accounted for the majority of serious workers' compensation claims (58%), while women represented 42%; however, the share of serious claims made by women has increased by 5.9% points over the past decade, reflecting rising female workforce participation. When adjusting for hours worked, men recorded a serious claims frequency rate 9.1% higher than women (7.0 versus 6.4 claims per million hours worked). The types of injuries also differ: women experience a higher proportion of serious claims for mental health conditions (17.2% compared with 8.2% for men), while men have higher proportions of physical injuries such as wounds, lacerations, amputations, and internal organ damage (17.7% versus 9.9%). Despite men receiving higher median compensation payments ($17,600 compared with $14,600), women experience slightly longer recovery periods, with a higher median time lost of 7.7 working weeks compared to 7.2 weeks for men.

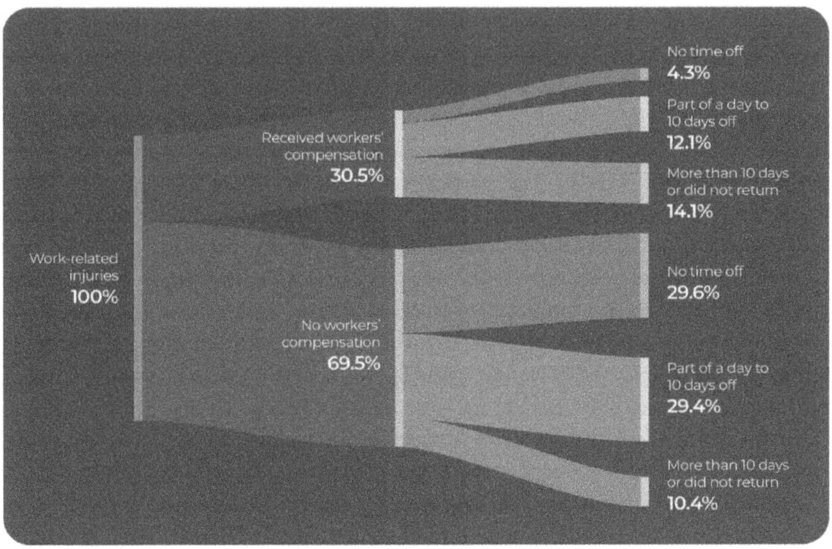

FIGURE 3.3 Work-related Injuries by compensation status and time off work, 2024.

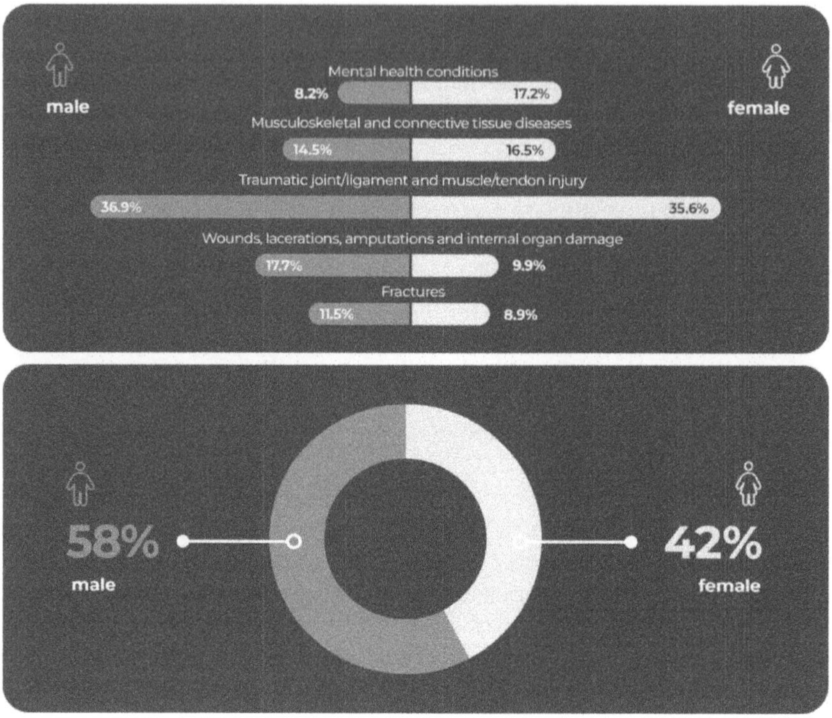

FIGURE 3.4 Work-related injury and illness: demographic patterns by sex and age, 2023–24.

Age is also an important factor in serious work-related injury and illness outcomes. Claim frequency generally increases with age, with the highest rates observed among workers aged 55–64 (9.5 serious claims per million hours worked) and those aged 65 years and over (10.0). In contrast, the lowest claim frequency rates (5.8 serious claims per million hours worked) were recorded for workers aged 25–34 and 35–44. Over the decade to 2023–24, the proportion of claims made by older workers has continued to rise, consistent with Australia's aging workforce. Older workers tend to require more time off and higher compensation when serious injuries occur, despite being less likely to experience overall injury or illness. Additionally, sex differences shift with age: men have higher claim frequency rates in younger age groups; however, from 55 years and over, women exhibit higher rates than men. These patterns highlight the interplay between age, gender, work exposure, and injury severity in shaping WHS outcomes across the Australian labor force.

Taken together, the four figures presented in this section demonstrate the complex and evolving landscape of work-related harm in Australia and highlight why ethically governed safety innovation is increasingly essential. Long-term fatality trends show progress but also reveal a recent plateau, while the 2024 fatality mechanisms data confirm that vehicle-related incidents and high-exposure occupations remain dominant sources of catastrophic harm. Similarly, the compensation pathways and demographic patterns show that injury severity, recovery time, and risk exposure are unevenly distributed across workers, occupations, age groups, and sexes. These empirical patterns illustrate both the need and the opportunity for emerging technologies, such as AI-enabled sensing, predictive analytics, automation, robotics, wearables, and digital twins, to reduce exposure to high-risk tasks, detect weak signals earlier, and strengthen real-time decision-making. However, we argue so far in this chapter that such innovations must be guided by a technological ethos that embeds dignity, fairness, transparency, and accountability into design and deployment. The data in Figures 3.1, 3.2, 3.3, and 3.4 reinforce this point: without ethical governance, technologies risk amplifying inequities, introducing new psychosocial harms, or creating opaque forms of surveillance and control. Thus, the quantitative evidence presented here provides not only situational awareness but also a practical foundation for the philosophically grounded and ethically aligned safety innovation model developed in the next chapter.

METRICS FOR ETHICALLY GOVERNED SAFETY INNOVATION (WHAT TO TRACK)

A practical approach to embedding ethics into safety innovation requires clear, measurable indicators that make ethical performance visible, trackable, and actionable over time. To support consistent governance and continuous improvement, organizations should monitor metrics across four interconnected domains: ethical purpose and alignment, responsible design and deployment, workforce and stakeholder impact, and trust, transparency, and assurance. Together, these domains provide a balanced set of signals that help determine whether new technologies and safety initiatives are not only effective; however, also socially responsible, fair, and aligned with worker expectations and regulatory standards.

Safety Performance (Traditional + Leading)

- TRIFR/LTIFR/OSHA recordables (as applicable)
- High-potential near misses (quality of reporting, not only frequency)
- Exposure metrics (time in high-energy zones, manual handling load)

Trustworthiness and Decision Quality (AI-Specific)

- False negative rate for critical hazards (missed high-risk events)
- Calibration (are confidence scores meaningful?)
- Drift indicators (performance changes over time)
- Override rates with justification (signals of mismatch or poor usability)

Ethical Safeguards (Governance Indicators)

- Proportion of data elements justified by purpose (data minimization score)
- Access logging compliance (% of access events recorded and reviewed)

- Contestability: time-to-resolution for worker challenges of tech outputs
- Audit completion rate (bias/robustness/security checks delivered on schedule)

Human Wellbeing (Safety 5.0 Alignment)

- Psychosocial harm indicators (workload, stress claims, fatigue signals)
- Perceived fairness and trust (survey items)
- Skills retention (do workers still demonstrate hazard recognition without the tool?)

This focus aligns with Safety 5.0 concerns that threats may shift toward stress and human-factor tensions as AI and cobots expand.

IMPLEMENTATION GUIDANCE: GOVERNANCE THAT PREVENTS "ETHICS WASHING"

Ethical integration fails when it is treated as branding. The following controls make ethics operational.

Assign Clear Accountability

- **Executive Owner:** accountable for ethos (what the tech is for; what it must never be used for).
- **HSE Owner:** accountable for safety assurance and learning integration.
- **IT/Security:** accountable for cyber controls and data integrity.
- **Worker Representatives/HSRs:** accountable for participation and contestability pathways.

Procurement Requirements (Put Ethics into Contracts)

Include contractual requirements for:

- data ownership and deletion
- audit rights and transparency artifacts (model cards, validation summaries)
- incident reporting support
- limits on secondary data use

9.3 Design Against Automation Bias

Because automation bias and complacency can persist even among skilled users [4], countermeasures should be built into the system:

- show uncertainty and rationale
- require cross-checks for high-consequence actions

- prevent "single-click" acceptance for critical decisions
- train for *trust calibration* (when not to rely on the tool)

Align with Trustworthy AI Requirements

Use a checklist aligned to trustworthy AI: human oversight, robustness/safety, privacy/data governance, transparency, fairness, accountability [5]. Treat it like a safety verification checklist, not a one-off statement.

CONCLUSION

Safety innovation is not ethically neutral. AI, automation, wearables, and digital twins can materially reduce injuries and improve early detection; evidence shows meaningful reductions in injury rates in some contexts and strong reported gains from wearable-enabled programs [6–8]. Yet these technologies can also introduce new harms through opacity, bias, surveillance, psychosocial stress, and automation bias. Ethical integration is therefore a core safety function, not an optional extra.

A strong technological ethos treats technology as a means to human well-being and resilient learning, consistent with the human-centric direction of Industry 5.0 and Safety 5.0. With ETHOS-SAFE, organizations can integrate emerging technologies through a lifecycle approach that produces concrete governance outputs, measurable indicators, and worker-centered safeguards. Ethical safety innovation, done well, does not slow progress; it prevents costly failure, protects dignity, and sustains trust, without which safety systems eventually degrade.

REFERENCES

1. Candrian, C. and Scherer, A., 2022. Rise of the machines: Delegating decisions to autonomous AI. *Computers in Human Behavior*, 134, p.107308.
2. Haynes, S., 2025. 5 The role of AI in human-centric risk decision-making. In *Operationalizing Twenty-First Century Safety: A Humancentric Practical Guide.*
3. Bowie, N.E., 2005. Kantian ethical thought. In *The Ethics of Human Resources and Industrial Relations*, pp.61–87.
4. Mosier, K.L. and Skitka, L.J., 2018. Human decision makers and automated decision aids: Made for each other?. In *Automation and Human Performance* (pp. 201–220). CRC Press.
5. Guillen-Aguinaga, M., Aguinaga-Ontoso, E., Guillen-Aguinaga, L., Guillen-Grima, F. and Aguinaga-Ontoso, I., 2025. Data quality in the age of AI: A review of governance, ethics, and the FAIR principles. *Data*, 10(12), p.201.
6. Yazdi, M., 2025. Safety through stories: Leveraging literature's lessons for practical safety insights. In *Safety-Centric Operations Research: Innovations and Integrative Approaches: A Multidisciplinary Approach to Managing Risk in Complex Systems* (pp. 51–72). Cham: Springer Nature Switzerland.
7. Matthias, B. and Reisinger, T., 2016, June. Example application of ISO/TS 15066 to a collaborative assembly scenario. In *Proceedings of ISR 2016: 47st International Symposium on Robotics* (pp. 1–5). VDE.

8. Patel, V., Chesmore, A., Legner, C.M. and Pandey, S., 2022. Trends in workplace wearable technologies and connected-worker solutions for next-generation occupational safety, health, and productivity. *Advanced Intelligent Systems*, 4(1), p.2100099.
9. Paccoud, I., Leist, A.K., Schwaninger, I., van Kessel, R. and Klucken, J., 2024. Socio-ethical challenges and opportunities for advancing diversity, equity, and inclusion in digital medicine. *Digital Health*, 10, p.20552076241277705.
10. Barrett, A.M., Hendrycks, D., Newman, J. and Nonnecke, B., 2022. Actionable guidance for high-consequence AI risk management: Towards standards addressing AI catastrophic risks. arXiv preprint arXiv:2206.08966.

4 The "Banality of Evil" in Workplace Health and Safety

INTRODUCTION

In the field of workplace health and safety, catastrophic failures often have humble origins. Major accidents and ethical breaches are frequently the culmination of numerous small acts of complacency or corner-cutting that went unchallenged. This phenomenon echoes what philosopher Hannah Arendt famously termed the "banality of evil" [1], describing how ordinary people can commit harmful acts unthinkingly, by dutifully following orders or adhering to routine without critical judgment. Arendt's insight emerged from her analysis of Adolf Eichmann, not as a diabolical villain, however, as a banal bureaucrat who caused enormous harm simply by "doing his job" and deferring responsibility upward [2]. In a parallel way, a "banality of evil" in the workplace might manifest as employees engaging in unsafe practices or unethical shortcuts because *"that's how it's always done"* or *"I was just following instructions."*

One subtle; however, telling indicator of a broken safety culture is the phrase: "I was told to do it." When this becomes the default excuse for unsafe actions, it signifies that personal responsibility has been supplanted by blind obedience. Such an environment allows risk to be normalized and perpetuated because individuals feel absolved of moral agency. The organization learns that keeping your head down and complying is safer (for one's job) than speaking up. Indeed, psychological studies on obedience found that ordinary people are disturbingly willing to inflict harm when they believe an authority will take responsibility for the consequences. In Stanley Milgram's classic experiments on obedience [3], 65% of participants delivered what they thought were lethal electric shocks to an innocent person simply because an authority figure insisted. Many subjects later explained their compliance with variations of *"I had no choice, the experimenter told me to continue,"* highlighting how diffusion of responsibility enables unethical or unsafe behavior.

Crucially, such dynamics are not confined to psychology labs or historical atrocities; they permeate everyday work life. In many industries, employees face pressure (implicit or explicit) [4] to *"get the job done"* even if it means bending safety rules. A rushed decision here, a shortcut there, a "just this once" exception that slowly becomes routine practice—over time, these deviations become the norm, a

DOI: 10.1201/9781003634720-4

process sociologists called the "normalization of deviance" [5]. Vaughan's study of the 1986 Challenger Space Shuttle disaster showed how NASA engineers and managers gradually accepted increasing risk (faulty O-ring seals) as normal since no immediate catastrophe ensued [6]. In her words, *"a clearly unsafe practice becomes considered normal if it doesn't immediately cause a catastrophe.*" This cultural drift allowed early warning signs to be misinterpreted, ignored, or missed completely until the day of the tragedy. The Challenger case is a vivid example of how "harm at work doesn't always start with something dramatic"; it often "begins quietly," with incremental departures from safety that accumulate under the radar. By the time a disaster strikes, those involved have long since rationalized the deviant practices as business as usual.

In this chapter, we delve into how such *banal* evil takes root and what can be done to counter it. We first connect Arendt's philosophical ideas to modern organizational contexts, examining how bureaucracy, social conformity, and thoughtlessness can erode ethical vigilance. We then discuss the role of leadership and hierarchy in either enabling or preventing these conditions. When leaders, through policies, incentives, or reactions, implicitly reward speed over safety, or punish those who raise concerns, they broadcast a powerful message that can permeate the entire organization. Over time, the workforce "learns the wrong lesson": that it is safer to stay silent and keep working than to question a dangerous situation. The result is an insidious culture of compliance and fear, in which risks are swept under the rug until they manifest in injury or catastrophe.

Yet, our analysis is not purely systemic; individual agency and ethics still matter. We consider cases of intentionally unsafe behavior or rule violations by employees and argue for a balanced approach to accountability. While reckless individuals must be held responsible for their choices, we must also ask: *What signals, pressures, or norms made that choice appear acceptable or necessary at the time?* By addressing these root causes, the organizational pressures and thought patterns that normalize risk, leaders can transform a culture of harm into one of harmony and safety. Drawing on both scientific research and philosophical teachings, we outline how organizations can cultivate a reflective, "thinking" culture that resists the banality of evil. This includes fostering open communication (so that *"just following orders"* is no excuse for silence in the face of danger), encouraging ethical deliberation at all levels, and designing incentives that align safety with success. In essence, we explore what it means to build a workplace where doing the right thing is as habitual as the behaviors we currently lament, a culture where safety is not seen as a hurdle to overcome; rather, as a core value woven into daily practice.

THE BANALITY OF EVIL IN ORGANIZATIONAL CONTEXT

Hannah Arendt coined the phrase "the banality of evil" in her 1963 report on the trial of Nazi war criminal Adolf Eichmann. Far from portraying Eichmann as a monstrous psychopath, Arendt was struck by his very ordinariness; he appeared to be a mundane, unimaginative bureaucrat dutifully following orders rather than zealously evil. The horrifying implication was that great evil can be committed by average people acting within flawed systems, simply by *not thinking* critically about the moral consequences of their actions. Arendt attributed Eichmann's crimes to a kind of "thoughtlessness," a disengagement from reality and inability to reflect from the standpoint of others. *In her view, this absence of critical thought and moral consideration is what rendered Eichmann's evil* "banal."

It was not monstrous passion, however, a routinization of duties and obedience, that enabled atrocities on a vast scale.

In modern workplaces, we thankfully deal with far less harms than genocide. However, Arendt's insights resonate in the patterns of everyday unethical or unsafe practices. Organizational scholars have applied Arendt's thesis to contemporary corporate wrongdoing, noting that wrongdoing often becomes "banalized" through bureaucratic processes, social normalization, and cognitive framing. Hernandez and Araos (2024) argue that many organizations inadvertently cultivate *thoughtlessness* in employees, discouraging critical thinking and personal responsibility through rigid routines and hierarchical pressures [7]. In corporate scandals or safety lapses, investigations frequently uncover that individuals "were just doing their job" or following standard procedure, yet the outcomes were harmful. It is precisely this contrast between the ordinariness of the actions and the enormity of the harm that defines banal evil.

Three interrelated mechanisms contribute to this phenomenon in organizations: bureaucratic obedience, social conformity, and diffusion of meaning. First, bureaucracy can create an environment where rules and orders are followed mechanically, without space for moral inquiry. People focus on fulfilling their narrow role (their *"slice"* of the task), and as a result, no one sees or feels responsible for the bigger picture. Safety responsibilities, in particular, can become so fragmented that everyone assumes someone else is handling the risk assessment. This was illustrated in the NASA Challenger case, where the decision to launch despite O-ring concerns passed through multiple management layers, each treating it as a technical waiver or paperwork exercise, rather than a life-and-death ethical choice. The bureaucratic mindset, Arendt observed, replaces personal accountability with adherence to process. If a danger is not explicitly flagged in the procedure, workers may carry on "by the book" straight into disaster.

Second, interactive socialization and group norms strongly shape behavior. New employees learn "how things are really done" by observing peers and superiors. If cutting corners or ignoring minor safety rules is a common practice, newcomers will adopt these deviant norms to fit in, often without realizing it. This is akin to Diane Vaughan's "normalization of deviance," where each small deviation from the rules, if unpunished, becomes accepted as a new normal. Over time a culture develops in which deviance is not seen as deviant at all. For example, in some industrial workplaces, not wearing certain protective gear or rushing through safety checklists might become customary, especially if those who take time for safety are teased or pressured. The collective mindset may be that "we have done it this way a hundred times and nothing bad happened." Such social proof powerfully dulls individuals' sense of risk and personal responsibility. The banal evil here lies in the quiet peer pressure that encourages everyone to keep their head down and go along, the unspoken "code of silence" that holds the flawed status quo in place.

Finally, cognitive framing and meaning-making play a role. Workers and managers might reframe or rationalize unsafe practices as acceptable through euphemisms and justifications. They tell themselves they are being "efficient," "practical," or "team players," rather than admitting they are compromising on safety. Psychological processes like moral disengagement enable well-intentioned people to participate in harmful activities by minimizing personal responsibility or harm, e.g., *"It's not really that dangerous,"* or *"This is above my pay grade to worry about."* Arendt emphasized how ordinary individuals can commit evils by losing the ability to think from the perspective of others and to ask about the true consequences. In workplaces, if employees are conditioned to think only about metrics, output, and pleasing the boss, they may stop asking the crucial question: *"Could this action hurt someone?"* The ultimate banality is when language and routine dull our moral imagination, paperwork replaces ethics, and we become, as Arendt feared, *functionaries* who simply do what we are told without reflection.

Table 4.1 below illustrates how prevalent "human error," often a proxy for these individual and cultural factors, is in workplace accidents across various industries. Notably, investigations often find that what is labeled *human error* (the immediate act, like pressing a wrong button or failing to secure a harness) has deeper organizational roots in training, culture, or systems design. In Arendt's terms, *thoughtlessness* is cultivated by the environment. When viewed superficially, however, it appears that a normal person just "made a mistake," another face of banal, everyday origins of harm.

Human error is frequently cited as a leading cause of accidents across industries, illustrating how often harm stems from the actions or inactions of individuals. However, contemporary safety science urges caution with the term "human error"; rather than an isolated personal failing, human errors are usually "symptoms" of

TABLE 4.1
Examples of Human Error Prevalence in Workplace Accidents

Sector	% of Accidents Attributed to Human Error	Source
All workplaces (general)	~80–90% of accidents involve human error factors [8]	OSHA data (via OSH)
Aviation (pilots)	~70–80% of aviation accidents are primarily due to human error [9]	NTSB/NASA studies; Vaughan (1996)
Construction	Up to 95% of injuries and incidents are attributed to worker error [10]	Industry safety reports (Australia)
Mining	~80% of mining sector incidents involve human error [10]	Queensland mining study (508 incidents)
Trucking/Transport	~80% of commercial truck crashes caused by driver error [10]	Transportation safety statistics

deeper issues in the system or culture. As psychologist James Reason famously noted, *"human error is not a cause; however, a consequence,"* provoked by upstream organizational factors and conditions [11]. In other words, if 90% of accidents are blamed on workers, one must ask: What aspects of training, environment, leadership, or design led those workers to slip up 90% of the time? The data in Table 1 underscore the need to examine how normal, well-intentioned employees can end up at the center of accidents, often because the *banal conditions* of their work made the mishap more likely (e.g., time pressure, fatigue, ambiguous procedures, or a culture of silence).

"I WAS TOLD TO DO IT": OBEDIENCE, AUTHORITY, AND NORMALIZED RISK

One of the most unnerving facets of workplace accidents and misconduct is how often those responsible offer the explanation: *"I was just doing what I was told."* This sentiment, sometimes called the "Nuremberg defense" (after the post-WWII trials) [12] or simply *obedience to authority*, is a hallmark of the banality of evil. In the context of occupational safety, it signals that employees have learned to prioritize compliance with orders over their own judgment, even when those orders entail risk. This deference can be fatal. If a supervisor instructs a worker to enter an unsafe area without proper lockout, or to run a machine beyond its safety limits to meet a deadline, an unquestioning *"Yes, boss"* can set the stage for injury. The worker's mindset becomes, *"It's not my role to question; my job is to follow orders."*

Decades of research in social psychology support how strong this impulse can be. Milgram's obedience experiments in the 1960s [13] demonstrated that ordinary individuals will inflict pain on another person when an authority figure insists, even against the individuals' own conscience. Participants who delivered the highest shocks often showed signs of extreme stress (sweating, trembling) yet continued after being prodded by the authority, later rationalizing that they *"had no choice"* since the experimenter was responsible. The key lesson was chilling: people can commit harmful acts more easily when they believe responsibility lies with someone above them. In a work setting, this dynamic means that if management implicitly or explicitly directs employees to take unsafe actions (or if the employees *perceive* that expectation), many will comply, especially if they fear consequences for refusal. The statement "I was told to do it" thus encapsulates both an excuse and a symptom; it excuses the individual (*I'm not personally culpable*) and symptoms an unhealthy culture (*I don't feel empowered to refuse dangerous directives*).

Organizational accidents often reveal a chain of such obedience. For instance, analyses of industrial disasters like the BP Texas City refinery explosion (2005) or the Deepwater Horizon oil rig blowout (2010) found that workers on the ground had misgivings about certain shortcuts; however, they went along with instructions due to production pressure and fear of reprisal [14]. Similarly, in the Boeing 737 MAX crisis (2018–2019) [15], internal emails later showed some engineers were aware of safety problems in the new flight control software; however, they did not forcefully escalate their concerns. Boeing's management had created an atmosphere that prioritized cost-cutting and speed over thorough safety checks to the point that employees remained silent or acquiescent about critical flaws. The tragic result, two crashes killing 346 people, illustrates how obedience and normalized deviance at multiple levels (design engineers, test pilots, regulators) led to a catastrophic "system accident." Each person in the chain did *just enough* to fulfill their immediate role, perhaps assuming others had validated the safety. In hindsight, it is easy to say someone should have blown the whistle. In practice, when *"faster, cheaper"* is the mantra from the top, those who hesitate or object face subtle or overt pressures: *"Don't be the obstacle; figure out how to deliver."* Thus, workers come to see risky shortcuts as necessary, even virtuous, to keep their jobs or please the boss.

Such pressures are alarmingly common. A recent survey of 1,000 US workers found that nearly one in four employees (25%) have accepted dangerous job responsibilities because they feared being fired if they refused [16]. In the same study, nearly 20% of workers said they feel uncomfortable raising safety concerns due to fear of retaliation or being ignored. This data paints a picture of fear-driven silence: many people will literally put themselves in harm's way, or allow hazards to go unreported, because they worry about the career consequences of speaking up. When asked why they took risks, respondents often cited financial or job security pressures; *"I couldn't afford to say no"* was a common refrain. Especially among lower-income workers, the economic stakes of losing a job can far outweigh the abstract risk of an accident, leading 27% of those earning under $50k to accept dangerous tasks they felt they had to shoulder. In effect, the workplace can become an environment where the short-term fear of the boss eclipses the longer-term fear of injury.

Figure 4.1 presents the top reasons cited by American workers for why they sometimes risk their safety at work. A 2025 survey revealed that the most common reason

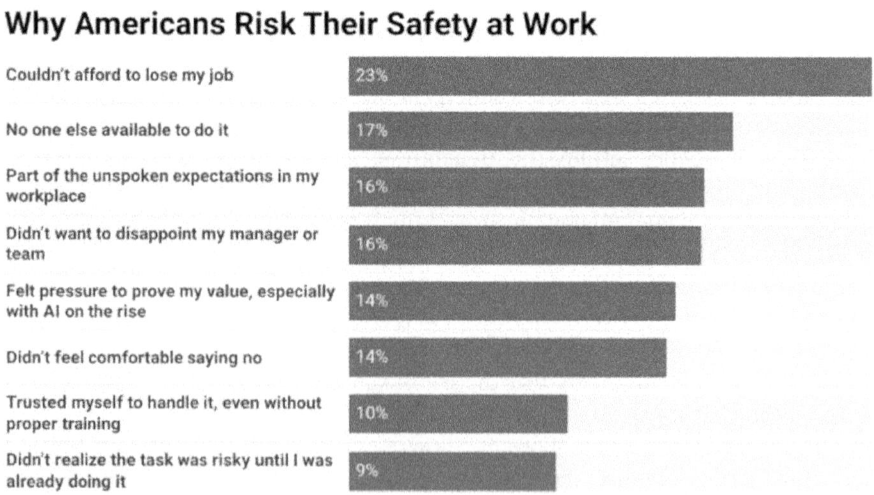

FIGURE 4.1 Key reasons American workers risk their safety at work (2025 survey): the role of job insecurity, workplace expectations, and systemic pressure.

(23% of respondents) was "I couldn't afford to lose my job," indicating financial/job insecurity as a driving force. Other frequent reasons included "No one else could do it" (17%) and "It was just expected as part of the job" (16%), as well as not wanting to disappoint managers or teammates (16%). A significant number (14%) also admitted they did not feel comfortable refusing a risky task or felt pressure to prove their value (often in competition with automation/AI) [17]. These findings underscore how systemic pressures and cultural expectations can push individuals to override safety concerns. Workers may internalize the idea that saying "no" is not truly an option, a clear parallel to the obedience phenomenon identified by Milgram, now playing out in everyday workplaces.

The phrase *"I was told to do it"* often surfaces after an incident as part of employees' testimony or in accident investigations. Rather than seeing it as a simple abdication of personal responsibility, organizations should treat it as a red flag indicating deeper cultural pathologies. It signals a lack of psychological safety, the term coined by Amy Edmondson to describe an environment where individuals feel free to speak up about problems or mistakes without fear of punishment. In a healthy safety culture, an employee who is instructed to do something hazardous would feel empowered to say, *"No, that's unsafe, and we need to handle this differently,"* or at least to question the order. Conversely, when psychological safety is low, the workforce reverts to a compliance mentality, and any resistance is stifled. This is how dangerous "groupthink" can set in, as was partly the case in the Challenger disaster: engineers at one point did raise alarms about the O-rings; however, management (under schedule pressure) discounted their input, and the engineers then acquiesced to the final launch decision. Afterwards, some might have thought, *"Well, I did voice concern; however, I was overruled, I did what I was told."*

To break this cycle, leadership must not only permit; however, actively encourage dissent when it comes to safety. As we discuss in the next section, the solution lies in leadership modeling and rewarding a mindset that *questions* rather than one that blindly complies. The "uncomfortable truth," as the introduction notes, is that safety is built or broken at the top of the hierarchy. It is leaders who set the tone for whether *"I was just following orders"* is viewed as a feeble excuse or an unacceptable cop-out. In a truly safety-conscious organization, no one should be able to hide behind orders, because orders that endanger are never given, and if they were, employees would be expected and empowered to challenge them.

LEADERSHIP, CULTURE, AND THE NORMALIZATION OF RISK

Organizational culture, "how we do things here," flows downward. Leaders, through their decisions and behaviors, are constantly teaching their employees what is really important. When it comes to safety, employees take cues from management's priorities. If workers see that meeting production targets at all costs is the boss's mantra, they will conclude that safety can be sacrificed for efficiency. If they see that every accident is blamed on a worker's carelessness (rather than examined for systemic fixes), they learn to hide mistakes and avoid reporting issues. Conversely, if management consistently demonstrates that *no* deadline justifies risking an injury, people will err on the side of caution. As one safety aphorism puts it, *"The behavior you reward (or tolerate) is the behavior you get"*.

Years of research underscore that leadership commitment is perhaps the single most critical factor in safety performance. High-reliability organizations (like in aviation or nuclear power) tend to have leaders who are almost obsessed with safety, conducting routine safety audits, encouraging open dialogue about near-misses, and investing in safety training even when times are tough. In broader industry, studies have found positive correlations between senior management's visible commitment to safety and lower accident rates, as well as higher job satisfaction and trust. Employees interpret management's *"say-do gap"* very keenly: If there is a safety policy that says "Safety First," however, workers observe that managers actually push for output while ignoring unsafe practices, the declared values mean nothing. In fact, hypocritical signals can breed cynicism, leading workers to take even more risks while rolling their eyes at the token safety slogans. Therefore, closing the gap between what leaders proclaim and what they enforce is crucial. Behavioral integrity, aligning actions with words, is needed to build a genuine safety culture.

One particularly damaging leadership failure is when managers react poorly to employees raising safety concerns. If a worker who brings up a hazard is met with irritation, dismissal, or worse, retaliation, a loud message is sent to everyone: *keep your mouth shut*. Even a subtle negative reaction, a sigh, a comment implying the person is slowing things down, can discourage future reporting. Over time, this creates a "chilling effect" where problems go unspoken. The organization becomes like a pressure cooker with a sealed lid: small issues and near-misses that could have been addressed continue to build until a serious incident forces attention. In the earlier mentioned survey, one in five workers feared retaliation or being ignored if they voiced safety issues [18]. This is a direct indictment of leadership climates in many workplaces. The antidote is for leaders to actively reward and thank employees for speaking up, even if the concern turns out minor or the suggested stop-work causes delay. Some companies have instituted formal programs to encourage reporting, such as *"Good Catch"* awards for spotting a potential hazard. The core idea is to counteract the natural fear of rocking the boat by making heroes of those who raise the alarm, not villains.

Another leadership influence is through goal-setting and incentives. If bonuses and promotions are tied strictly to productivity metrics with no regard for safety, people will get the message that safety is a secondary consideration or even an obstacle. A classic example is when a company boasts of "X million hours without a lost-time injury" and rewards teams for hitting injury-free milestones. While the intent is to promote safety, such incentives sometimes backfire by encouraging underreporting; nobody wants to be the one who "breaks the streak," so injuries and incidents get swept under the rug. A more insidious case is quotas or performance targets that are unrealistic without cutting safety corners. For instance, if a delivery company's dispatch schedule effectively forces drivers to speed or skip breaks to meet targets, management is tacitly endorsing risk-taking. Workers adapt their behavior to what they perceive leadership truly values. In this sense, leadership may *indirectly create "intentional" unsafe behavior* by shaping the context in which it occurs. A worker who knowingly bypasses a safety device to speed up work might be doing so because past behavior of management signaled that speed is king, and the only feedback they've received is on productivity, not safe procedure.

To illustrate how leadership behaviors translate into safety outcomes, Table 4.2 provides a few scenario examples of leadership actions and the cultural messages they impart. These examples demonstrate how seemingly routine management practices can make the difference between a vigilant safety culture and a complacent, high-risk one:

These examples show how leadership actions trickle down to influence safety culture. Leaders send signals in every decision, whether they realize it or not. Consistent,

TABLE 4.2
Leadership Behavior and Its Impact on Safety Culture

Leadership Behavior or Decision	Message Conveyed to Employees	Likely Impact on Safety Culture
Manager openly ignores rules (e.g., skips safety goggles on site visits or bypasses lockout procedures to save time).	"Even the boss doesn't follow the rules, so they must not be that important."	Erosion of rule adherence. Employees mimic the non-compliance, normalizing rule-breaking as acceptable.
Supervisor pressures for output after an employee raises a safety concern (e.g., *"Just get it done; we'll deal with that later"*).	"Production matters more than safety; raising concerns will be brushed off."	Decreased reporting of issues. Shortcuts proliferate as workers believe management prefers ignorance to delays.
After an incident, upper management blames an individual publicly (fires or disciplines the front-line worker involved, with no examination of system factors).	"If you get hurt or make a mistake, you'll be punished. Management sees accidents only as worker failures."	Emergence of a blame culture. Employees hide mistakes or injuries for fear of blame, and underlying causes (training gaps, workload issues) remain unaddressed.
Senior leader responds positively to bad news (e.g., thanks an employee for pointing out a major hazard or near-miss, even if it causes work stoppage).	"Honesty and safety come first. Management wants to know about problems and will support those who speak up."	Improved psychological safety. Employees more likely to report hazards and near-misses, enabling proactive fixes before accidents occur.
Company sets a safety performance goal (e.g., reducing incident rates) *alongside* production goals, and tracks both in management meetings.	"Safety is equally valued as productivity. Success means doing work safely, not just quickly."	Balanced priorities. Workers feel less pressure to cut corners; they see that doing a job safely is part of doing it well. This tends to correlate with lower accident rates and higher morale.

authentic leadership commitment to safety creates an environment where employees understand that safe behavior is expected and supported (last two examples). On the contrary, if leaders flout safety rules or penalize those who report issues, they inadvertently cultivate a culture of fear and corner-cutting (first three examples). Extensive research confirms that blame-oriented and fear-driven cultures are detrimental to safety: they destroy trust and openness, causing employees to adopt a *"cover-your-back"* mentality rather than a *"collective problem-solving"* mentality. In contrast, leaders who foster trust and learning from errors (a *"just culture"* approach) enable continuous improvement in safety outcomes. In sum, the tone at the top is a critical determinant of whether the "banal" risks in day-to-day work are kept in check or allowed to grow.

ACCOUNTABILITY AND ETHICAL REFLECTIONS: FROM HARM TO HARMONY

When unsafe behavior or an accident occurs, a natural question arises: *Who is to blame?* Traditionally, the answer often pinned fault on a proximal cause, the worker who erred or violated a rule. However, as we have explored, this "person approach" to blame can be simplistic. A more constructive question is: *Why did this happen and how can we prevent it?* This shifts the focus from solely individual culpability to organizational learning and systemic improvement. In the philosophy of safety management, this idea aligns with what James Reason and others advocate: a "just culture." A just culture seeks to balance accountability with a recognition that human mistakes are inevitable and usually induced by system factors. In a just culture, gross negligence or willful harmful acts (e.g., an individual coming to work intoxicated or intentionally bypassing safety for personal convenience) are not tolerated; these incur sanctions because they involve conscious choice. However, the vast majority of incidents that fall into the realm of "honest mistakes" or slips, the banal errors, are treated as opportunities to fix the system, not simply punish the individual.

To illustrate, consider a scenario: a lab technician skips a step in a safety protocol and is splashed with a chemical, suffering burns. In a blame culture, the incident report might conclude "technician failed to follow procedure, discipline accordingly." In a just culture, the investigation would dig deeper: *Why* was the step skipped? Was the procedure cumbersome or unclear? Was the technician rushing due to supervisors' pressure or understaffing? Had there been previous near-misses indicating that this step was commonly overlooked? By addressing these questions, the organization might find that the procedure needs revision for clarity, or that the technician's unit is chronically overworked, or that training on that protocol was insufficient. Systems thinking would lead to solutions (streamlining the procedure, adjusting workload, refreshing training) that benefit many, rather than simply terminating or scolding one "bad apple." This approach aligns with ethical frameworks that emphasize *learning and prevention over retribution*. It echoes philosopher John Dewey's view that accidents and mistakes should be seen as information for improving future practice, not as moral failings [19].

From a broader ethical standpoint, leaders must interrogate what pressures or incentives made an unsafe choice appear reasonable to the person at the time. This is a hard question because it may reveal leadership's own role in the chain of causation. Going back to the lab example, perhaps management set unrealistic throughput goals that tacitly encouraged technicians to skip steps for the sake of speed. Recognizing this is uncomfortable; however, it is precisely how organizations grow ethically: by taking responsibility at the leadership level, not only demanding it at the worker level. In essence, leaders should ask of any incident: *"In what way did our system set this person up to fail?"* This question reflects moral humility and accountability at the top, virtues that philosophers like Confucius or Plato might say are marks of just leadership. It also reflects an understanding of the banality of systemic evil, that often it is ordinary organizational decisions (to cut budgets, to push for faster output, to ignore complaints) that create the conditions for frontline workers' "errors."

That said, emphasizing systems does not mean absolving individuals of all agency. It is important to maintain that safety is everyone's responsibility, from the CEO to the newest hire. This is where ethical training and philosophical insight for individuals can complement systemic measures. For example, drawing from Kantian ethics, one could instill the idea that each person has a duty to treat safety as a categorical imperative: *Never treat the health and lives of yourself or co-workers merely as a means to an end (like profit or convenience); however, always as ends in themselves.* Such a principle encourages employees at all levels to speak out or refuse when they recognize a serious safety breach, even if an authority tells them otherwise. It appeals to personal conscience, saying in effect that *"just following orders"* is not an excuse if it means violating a fundamental moral duty to not cause harm. Similarly, virtue ethics would encourage the cultivation of courage and prudence in the workforce: courage to voice concerns or halt work when needed, and prudence to foresee potential harms. A worker who has internalized the virtue of responsibility will be less likely to hide behind instructions and more likely to act rightly, even under pressure.

Leaders, for their part, can draw on the philosophy of benevolence and justice. The ancient Chinese philosopher Confucius taught that a good leader is like a father/mother figure who cares for the well-being of the people; in a modern sense, this translates to genuinely caring about employees' safety and not merely the bottom line. Confucian thought would suggest that when leaders demonstrate *ren* (humaneness) and *yi* (righteousness), an ethos of loyalty and trust flourishes, and employees reciprocate by working diligently and also by watching out for each other. In practical terms, a leader practicing these values might implement an open-door policy for safety issues, devote resources to ergonomic improvements out of concern for workers' health, or personally intervene to stop hazardous operations until improvements are made. Such actions build harmony in the workplace, not the false harmony of silence and coerced obedience; however, the genuine harmony of aligned values, where everyone from top to bottom shares the commitment to safety as a core value.

In moving "from harm to harmony," education and dialogue are key. Ethics training should not be a dry compliance module; however, it should be an engaging discussion of real dilemmas employees might face: e.g., *"What would you do if your supervisor asks you to skip a safety step to meet a deadline?"* Role-playing these scenarios and examining them under different ethical lenses (duty, consequences, virtue) can prepare workers to handle pressure in the moment. It can also surface disconnects in policy; for instance, if employees feel they *cannot* refuse such a request without retaliation, that needs to be addressed proactively by management. Some companies incorporate philosophy or ethics discussions into their safety meetings, reinforcing that safety is about doing the right thing, not just following rules. By highlighting stories of employees who did speak up and prevented harm (and rewarding them), organizations reinforce the narrative that *"the first duty is to safety"*.

On a structural level, whistleblower protections and independent reporting channels are important for breaking the "conspiracy of silence" that banal evil relies on. Employees must have confidence that if they raise a serious concern, even going outside the chain of command if necessary, they will be protected. Strong whistleblower laws and internal policies reflect a societal acknowledgment of the banality of evil: we recognize that pressures can make good people stay quiet, so we create systems to counteract that, giving voice to conscience. For example, the US Occupational Safety and Health Administration (OSHA) has provisions protecting workers who refuse unsafe work or report violations. Enforcing these and making employees aware of them is part of cultivating an ethical, safety-first climate.

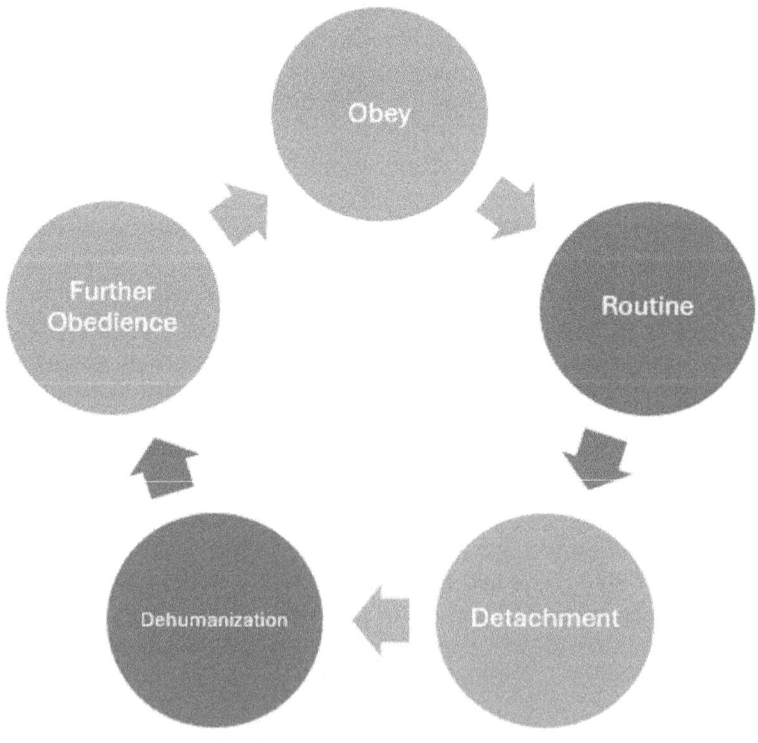

FIGURE 4.2 The "Moral Disengagement Loop"

In summary, preventing harm and fostering harmony in workplace safety requires aligning both the system and the individual mindset with ethical principles. Philosophers teach us that evil (or harm) often stems not from malice; however, from *lack of reflection*, and thus the antidote lies in *thoughtfulness*. Arendt would likely counsel organizations to nurture *thinking employees* at every level: people who are empowered and expected to stop and think, "What am I actually doing? What might the consequences be? Does this feel right?" Such mindfulness can interrupt the auto-pilot mode that banality thrives on. It can transform an order-following culture into a questioning culture, not in a rebellious way, however, in a responsibly skeptical way that catches errors and ethical issues early. To achieve this, leaders must lead by example, demonstrating that raising concerns is not insubordination, however, integrity.

To conclude, as it can be seen from Figure 1, the cycle of the banality of evil emerges through a gradual and often unnoticed progression in which ordinary individuals become participants in harmful systems. The diagram illustrates how the process begins with simple obedience, which then transforms into routine actions performed without critical reflection. This routinization leads to emotional and moral detachment, creating the psychological distance necessary for individuals to carry out tasks that may conflict with their personal values. As detachment deepens, dehumanization can occur, making it easier to view others as objects rather than people deserving of empathy or ethical consideration. Ultimately, this cycle reinforces itself through further obedience, perpetuating harmful behaviors not out of malice but through unexamined compliance and the normalization of destructive practices.

CONCLUSION

The "banality of evil" in workplace health and safety reminds us that significant harm often originates from ordinary decisions made by ordinary people in the course of doing their jobs. Small acts of negligence or unthinking obedience, repeated daily, can accumulate into organizational disasters. The anecdotes of workers saying *"I was told to do it"* after an accident encapsulate how diffuse responsibility and unquestioning hierarchy can normalize risk. This chapter has highlighted that the true locus of safety (or its absence) is in the mundane interactions and choices shaped by organizational culture. Leadership stands as the pivotal influence: when leaders implicitly reward unsafe speed or punish honest reporting, they seed a culture in which employees learn to trade safety for self-preservation or performance. Over time, this creates a vicious cycle of normalization, where everyone assumes someone else has vetted the risk or that "this is just how things are done," until the day an avoidable tragedy strikes.

Breaking out of this banality requires deliberate ethical leadership and an empowered workforce. It means instituting a culture where speaking up is valued over shutting up, where doing what is right is valued over doing what is told. The best organizations foster a sense of collective responsibility: workers don't say *"I was told to do it,"* they say *"I didn't feel it was safe, so I stopped and informed my* supervisor." And the supervisor, in a positive culture, thanks them and works to resolve the issue rather than scolding them. Achieving this state is not trivial; it often requires overcoming ingrained habits and economic pressures. However, the effort is worthwhile, not only to prevent harm; however, to create a more humane and engaged workplace.

Philosophically, what is at stake is respect for human dignity. Workplace safety is ultimately about valuing human life and well-being above other priorities. Philosophers from Immanuel Kant to modern ethicists would argue that treating workers as ends in themselves, not merely as means to production, is a moral imperative for any organization. In practical terms, this translates to safety being a non-negotiable value. When that ethos is genuinely in place, the banal evils of everyday corner-cutting and silence cannot easily take root because everyone from leadership to the front line is continually attentive and ready to intervene at the first signs of danger.

In closing, the journey "from harm to harmony" in workplace safety can be guided by a blend of scientific insight and philosophical wisdom. From science, we learn about the psychology of compliance, the sociology of deviance normalization, and the management practices that reduce accidents. From philosophy, we are reminded of the ethical dimensions, the importance of critical thinking, the courage to uphold good in the face of pressure, and the moral duty of those in power to care for those in their charge. By integrating these lessons, organizations can hope to banish the "banality" of unsafe practices and replace it with an actively conscientious safety culture. In such a culture, evil (or harm) does not get to hide in the ordinary; instead, *everyday good*, the simple, daily acts of paying attention, speaking up, and watching out for one another, become the norm.

REFERENCES

1. Borba, R. and Milani, T.M., 2017. The banality of evil. *Journal of Language and Discrimination*, 1(1), pp.7–33.
2. Norberg, J., 2010. The political theory of the cliché: Hannah Arendt reading Adolf Eichmann. *Cultural Critique*, (76), pp.74–97.
3. Slater, M., Antley, A., Davison, A., Swapp, D., Guger, C., Barker, C., Pistrang, N. and Sanchez-Vives, M.V., 2006. A virtual reprise of the Stanley Milgram obedience experiments. *PloS One*, 1(1), p.e39.
4. Caza, B.B., Tiedens, L. and Lee, F., 2011. Power becomes you: The effects of implicit and explicit power on the self. *Organizational Behavior and Human Decision Processes*, 114(1), pp.15–24.
5. Price, M.R. and Williams, T.C., 2018. When doing wrong feels so right: Normalization of deviance. *Journal of Patient Safety*, 14(1), pp.1–2.
6. Vaughan, D., 2004. Theorizing disaster: Analogy, historical ethnography, and the Challenger accident. *Ethnography*, 5(3), pp.315–347.
7. Gottschalk, P. and Hamerton, C., 2025. Domineering leadership and the challenge of conformance repair. In *Corporate Conformance and the Social License: A Criminological Perspective* (pp. 75–128). Cham: Springer Nature Switzerland.
8. Reason, J., 1995. Understanding adverse events: Human factors. *BMJ Quality & Safety*, 4(2), pp.80–89.
9. Kharoufah, H., Murray, J., Baxter, G. and Wild, G., 2018. A review of human factors causations in commercial air transport accidents and incidents: From to 2000–2016. *Progress in Aerospace Sciences*, 99, pp.1–13.

10. Yazdi, M., 2024. *Artificial Intelligence in Workplace Health and Safety: Data-Driven Technologies, Tools and Techniques.*(1st ed.) CRC Press. https://doi.org/10.1201/9781003515173
11. Tang, J., Bao, Y., Wang, L., Lu, H., Wang, Y., Guo, C., Liu, J. and Zhou, B., 2013, December. A Bayesian network approach for human reliability analysis of power system. In *2013 IEEE PES Asia-Pacific Power and Energy Engineering Conference (APPEEC)* (pp. 1–6). IEEE.
12. Wolfe, R., 1980. Putative threat to national security as a Nuremberg defense for Genocide. *The ANNALS of the American Academy of Political and Social Science*, 450(1), pp.46–67.
13. Milgram, S., 1963. Behavioral study of obedience. *The Journal of Abnormal and Social Psychology*, 67(4), p.371.
14. Holmstrom, D., Altamirano, F., Banks, J., Joseph, G., Kaszniak, M., Mackenzie, C., Shroff, R., Cohen, H. and Wallace, S., 2006. CSB investigation of the explosions and fire at the BP Texas City refinery on March 23, 2005. *Process Safety Progress*, 25(4), pp.345–349.
15. Thompson, D., 2024, August. Boeing 737 max crisis of 2018–2019: A content analysis. In *Proceedings of the 37th International RAIS Conference on Social Sciences and Humanities*, vol. 2 (pp. 55–70). Scientia Moralitas Research Institute.
16. Allen, D., Hines, E.W., Pazdernik, V., Konecny, L.T. and Breitenbach, E., 2018. Four-year review of presenteeism data among employees of a large United States health care system: A retrospective prevalence study. *Human Resources for Health*, 16(1), p.59.
17. Newton-Levinson, A., Griffin, K., Blake, S.C., Swartzendruber, A., Kramer, M. and Sales, J.M., 2024. "I probably have access, but I can't afford it": Expanding definitions of affordability in access to contraceptive services among people with low income in Georgia, USA. *BMC Health Services Research*, 24(1), p.709.
18. Cortina, L.M. and Magley, V.J., 2003. Raising voice, risking retaliation: Events following interpersonal mistreatment in the workplace. *Journal of Occupational Health Psychology*, 8(4), p.247.
19. Dewey, J., 1891. Moral theory and practice. *The International Journal of Ethics*, 1(2), pp.186–203.

5 Technological Ethos and Safety Innovations
Ethical Integration of Emerging Technologies

INTRODUCTION

Emerging technologies such as artificial intelligence (AI), robotics, and automation are rapidly transforming safety practices across industries. This intersection of technology and ethics forces a re-examination of how we define and uphold safety in the modern era. The prevailing technological *ethos* often emphasizes speed, efficiency, and innovation, encapsulated in Silicon Valley's old mantra of "move fast and break things" [1]. By contrast, traditional safety culture prioritizes caution, reliability, and the prevention of harm. This tension raises profound philosophical questions: How can we harness cutting-edge innovations to *improve* safety without "breaking" the ethical principles that protect people? In what ways do AI and automation challenge longstanding safety paradigms, and how can we ethically integrate these technologies into practice? This report explores these questions, examining how new technologies reshape safety thinking and proposing frameworks to align technological innovation with ethical safety outcomes.

TECHNOLOGY AND ETHICS IN SAFETY: A NEW FRONTIER

The convergence of advanced technology with safety management is more than a technical issue; it is inherently ethical. At its core, safety is a moral commitment to prevent harm and protect life. New technologies amplify both our capability to fulfill this commitment and the risks of falling short. Philosophically, we must balance a utilitarian impulse to reduce overall harm (e.g., by using automation to eliminate human error) with deontological principles that respect rights and transparency for individuals affected by these systems. The *technological ethos* driving innovation, favoring optimization, data-driven decision-making, and often corporate interests, must be reconciled with a *safety ethos* centered on human well-being and "do no harm."

This intersection is evident in debates around AI ethics. For example, business leaders are increasingly realizing that purely technical solutions are not enough; ethical foresight is needed to avoid worst-case scenarios that endanger people. New technology can produce *"ethical nightmares,"* from privacy violations to unsafe failures, if deployed recklessly. In the safety domain, this means engineers and ethicists

DOI: 10.1201/9781003634720-5

must work together. An algorithm might optimize productivity; however, if it undermines workers' dignity, privacy, or safety, it violates ethical standards. Thus, the integration of tech and safety is as much about values and governance as about sensors and software [2]. We need a *philosophical shift*: instead of treating ethics as a brake on innovation, view it as an essential guardrail that ensures innovations actually *enhance* human safety.

CHALLENGING AND RESHAPING TRADITIONAL SAFETY PARADIGMS

Innovations like AI and automation are profoundly challenging traditional safety paradigms. Classic safety management (sometimes termed "Safety-I") [3] views safety as the *absence of accidents*, focusing on preventing things from going wrong. It relies heavily on human oversight, checklists, and reactive measures when incidents occur. In contrast, today's technologies enable a proactive approach, real-time monitoring, predictive analytics, and autonomous interventions that can prevent incidents before they happen. This is akin to the emerging "Safety-II" paradigm, which defines safety as the system's ability to succeed under varying conditions. AI-driven systems can continuously learn from data and adjust, theoretically increasing resilience and catching subtle warning signs humans might miss.

Reshaping the Paradigm, Positive Impacts: In many cases, AI and automation have improved safety outcomes, validating the hope that technology can be a powerful safety tool. For instance, AI enables *real-time hazard detection and predictive risk assessment* in workplaces, marking a "paradigm shift" in occupational health and safety toward more preventive and data-driven practices. Wearable sensors and machine learning can predict when a worker is fatigued or a machine is about to fail, allowing interventions before an accident occurs. Table 5.1 illustrates how increased automation is statistically linked to lower injury rates, according to recent studies and forecasts:

As automation and AI are adopted (about 30% of tasks automated by 2030), significant drops in workplace injury rates are expected in major sectors. Nationally, injuries are projected to fall by ~5.9% (from 2.43 to 2.29 injuries per 100 workers), preventing an estimated 161,000 injuries per year by 2030. Historically, even a 10%

TABLE 5.1
Automation and Reduction in Workplace Injuries

Sector	Predicted Injury Rate Reduction by 2030	Fewer Injuries per Year (approx)
Private Healthcare	6.3%	~30,000 fewer injuries annually
Goods Manufacturing	5.5%	~18,143 fewer injuries annually
Retail Trade	4.9%	~16,405 fewer injuries annually

increase in automation has been associated with roughly a 2% reduction in injuries, evidence that automating dangerous tasks *can directly improve* safety.

Real-world examples bear this out. In the mining industry, which Australia has led in automating, removing human drivers from heavy vehicles has dramatically reduced certain accidents. Rio Tinto's autonomous haul trucks in the Pilbara mines led to a *40% reduction in safety incidents*, largely by eliminating human error in haulage operations. Similarly, Canada's Suncor mine saw a 35% drop in incidents after deploying autonomous trucks. These improvements reflect a new paradigm where machines handle the most hazardous tasks, protecting workers from exposure to danger. Automation also often enforces a consistent, rules-based approach; for example, a self-driving haul truck will never drive drunk or tired, nor take dangerous shortcuts, which contributes to safer outcomes.

New Risks and Paradigm Tensions: However, these same technologies also introduce new types of risks and ethical dilemmas, meaning the safety paradigm must expand rather than simply celebrate victory. A striking example comes from automated warehouses. Research on Amazon's robotic fulfillment centers found a dual effect: severe injuries (e.g., traumatic injuries causing lost workdays) *decreased by 40%* in warehouses with robots, a positive outcome; however, minor injuries (e.g., musculoskeletal strains) *increased by 77%* in those same high-tech facilities. The automation took over heavy lifting (reducing acute accidents); however, it simultaneously sped up the pace of work, leading to more repetitive strain and fatigue injuries for the human workers working alongside robots. This *"reshuffling of risk"* means the safety paradigm cannot just count fewer broken bones and declare success; it must also address subtler harms like ergonomic stress and mental burnout that technology may exacerbate. As one expert noted, simply introducing robots *"is not enough*"; without careful job redesign and realistic workload expectations, automation can create new health and safety hazards even as it solves others.

Furthermore, AI systems challenge the traditional paradigm of human oversight. In safety-critical decisions (for example, an autonomous vehicle making a split-second choice to swerve and risk one life to save another), ethical judgment is now partly delegated to machines [4]. This raises the classic "trolley problem" scenario: How should an AI be programmed to value lives in an accident? Such questions were previously academic; however, they become urgent when *automation directly controls outcomes.* Traditional safety engineering would avoid putting anyone in harm's way; now, AI might have to choose *who* to put in harm's way in a no-win scenario. Our existing safety frameworks struggle with this, requiring new ethical paradigms of accountability and decision-making under uncertainty.

Accountability and Transparency: Traditional safety practice relies on clear accountability; a human engineer, operator, or manager can be identified when procedures fail [5]. AI complicates this. If a self-driving car or an industrial AI system causes an accident, who is responsible—the designer, the user, the algorithm itself? This ambiguity challenges legal and ethical paradigms. It has prompted initiatives worldwide to require transparency and oversight of AI decisions. For example, the European Commission's guidelines for *"Trustworthy AI"* insist on human agency and oversight and emphasize robustness and safety to ensure AI systems *do not cause unintended harm.* Similarly, Australia's own AI Ethics Principles include *"Reliability and safety"* and *"Accountability"* as core tenets, calling for AI to operate as intended without unreasonable risks and for humans to be identifiable and responsible for AI outcomes. These evolving norms are pushing the safety paradigm from *"operator error"* models toward systemic accountability, where organizations must anticipate and mitigate AI-driven risks.

In summary, AI and automation are reshaping safety paradigms in a two-fold way: they enable a leap forward in preventing and minimizing certain risks, while also forcing us to broaden our concept of safety to include algorithmic behavior, data ethics, and human-tech interactions. The net result can be profoundly positive, fewer workers in mines and factories getting hurt or killed, but only if we adapt our safety philosophies to manage the new complexities that come with intelligent, autonomous systems.

ETHICAL CHALLENGES POSED BY EMERGING SAFETY TECHNOLOGIES

To integrate emerging technologies into safety practices ethically, we must directly address the ethical challenges and dilemmas they bring. Many of these challenges arise from the very features that make AI and automation powerful:

- **Data and Privacy:** Modern safety systems often rely on collecting vast data (e.g., wearable health monitors, location trackers on workers, CCTV with AI analysis) [6]. This raises privacy concerns. An AI-driven safety program might continuously monitor employees' movements and vital signs to prevent incidents; however, such surveillance can infringe on autonomy and dignity if not carefully governed. Ensuring *consent, privacy, and data protection* is paramount. For example, an AI that monitors stress levels via wearables must secure that personal data and use it only for legitimate safety purposes. Ethical integration means transparency about what data is collected and giving workers a say; otherwise, a well-intended safety measure could become an unacceptable invasion of privacy.
- **Bias and Fairness:** AI systems can inadvertently introduce bias in safety or risk assessments. If the data used to train an AI is skewed, the AI's decisions might unfairly favor or disfavor certain groups [7]. Imagine an AI that allocates safety resources or flags "risky" worker behavior; if it is trained on biased incident records, it might single out certain demographics unfairly. Ethically, we must ensure these systems are audited for fairness and do not become tools of discrimination. This ties into the broader AI ethics principle of *justice*: technology in safety should *protect everyone equitably*, not just those who fit the norm.
- **Loss of Human Control and Skills:** Automation can lead to humans becoming complacent or deskilled in critical safety tasks [8]. If pilots rely on autopilot, or plant operators on an AI alarm system, their own ability to

intervene might atrophy. The ethical concern is that over-reliance on automation could erode the human capability to ensure safety, especially in emergencies. Therefore, a balance is needed: keep humans *in the loop* for oversight and provide training so that people working with AI understand its limits and can step in when needed. This is reflected in ethics guidelines calling for *human-centered design*; AI should *augment* human decision-making, not replace it entirely in matters of safety.
- **Job Displacement and Psychological Impact:** A broader ethical issue is the impact of automation on employment and mental well-being. Replacing dangerous jobs with robots can save lives, but it can also eliminate livelihoods. Ethically integrating technology means mitigating the social impact, e.g., by retraining workers for new roles overseeing the technology, rather than simply rendering them obsolete. Even for workers who remain, working in a high-tech environment can bring stress and anxiety. Studies have noted that workers monitored by AI or working alongside intelligent systems often report heightened stress, a *"loss of control,"* and alienation. For instance, warehouse employees in automated facilities felt increased pressure to meet machine-paced performance metrics, leading to burnout. Safety is not just physical; psychological safety is also crucial. Thus, the ethical approach to tech integration calls for attention to mental health, reasonable workloads, and maintaining a sense of purpose for workers in an automated workplace.
- **Unintended Consequences and Complex Failures:** Complex tech systems can fail in unexpected ways. An autonomous safety system might encounter a scenario its designers never anticipated. The ethical challenge is how to foresee and prevent *"unknown unknowns."* This is where precautionary thinking and ongoing vigilance are needed. Techniques like rigorous simulation, scenario planning (including *worst-case ethical scenarios*), and continuous monitoring of AI decisions become part of the safety ethic. A notable example is the fatal accident of a self-driving test vehicle in 2018, which revealed that the AI failed to recognize a pedestrian in time, highlighting how *literal life-and-death bugs* can lurk in complex systems. Ethical integration requires humility and caution: acknowledging that new tech can and will fail and building fail-safes (both technical and organizational) to catch failures before they harm people.

In confronting these challenges, companies and regulators around the world are actively developing guidelines and laws. The European Union's AI Act [9] is a landmark effort that will impose strict requirements (like risk assessments, transparency, and human oversight) on "high-risk" AI systems involved in safety-critical functions. Australia, too, has moved to set explicit obligations for workplace AI safety. New South Wales in 2025 introduced the WHS Amendment (Digital Work Systems) Bill, the first of its kind in Australia, which will require employers to ensure workers are not put at risk by AI, algorithms, or automation in the workplace. This kind of

TABLE 5.2
The main WHS legislation

Topic	Refined Summary
Primary Duty of Care (s19 WHS Act)	The primary duty of care is amended to require that a PCBU ensures, *so far as is reasonably practicable*, that workers' health and safety is not put at risk by the use of digital work systems.
Definition: "Digital Work System"	Defined as an algorithm, artificial intelligence, automation, or online platform. The originally proposed reference to "software" has been removed to avoid unintentionally covering systems such as email or basic messaging tools.
Specific Duty—Digital Work Systems (s21A WHS Act)	A PCBU must ensure, *so far as is reasonably practicable*, that workers' health and safety is not put at risk from how a digital work system allocates work. A PCBU must consider whether the system results in: • excessive or unreasonable workloads • excessive or unreasonable metrics for performance assessment or tracking • excessive or unreasonable monitoring or surveillance • discriminatory practices or decision-making
WHS Permit Holders—Access to Digital Work Systems (s118 and s118A WHS Act)	A WHS entry permit holder may require a PCBU to provide reasonable assistance to access and inspect any digital work system relevant to a suspected WHS contravention. Access will be subject to regulator-issued guidelines (not yet published). A PCBU is not required to provide access if doing so would breach a Federal or State law. Penalties for refusal: • $66,770 for a corporation • $13,310 for an individual
Review Against Model WHS Laws (s276D WHS Act)	If Safe Work Australia's ongoing review of the model WHS laws results in provisions dealing with substantially the same subject matter, these amendments must undergo review. This may lead to further amendments to align with the model laws.

legislation is emerging with recognition that integrating tech and safety is not just best practice; however, it is a legal and ethical mandate, refers to Table 5.2.

FRAMEWORKS FOR ETHICALLY INTEGRATING NEW TECHNOLOGIES INTO SAFETY PRACTICES

To bridge the gap between technological potential and ethical practice, we need robust frameworks for ethical integration. Such frameworks provide structured guidance so that organizations can innovate *safely* and *responsibly.* Below, we propose key components and principles of an ethical integration framework, informed by global best practices and philosophical reflection:

1. **Ethical Risk Assessment and Foresight:** Before deploying an emerging technology, perform a thorough ethical risk analysis. This means identifying potential ways the technology could cause harm or violate ethical norms, even unintentionally. Companies should explicitly articulate worst-case scenarios ("ethical nightmares") and how they will prevent them. For example, if introducing an AI-driven machinery control system, consider: Could it malfunction and endanger workers? Could it make biased decisions that affect safety? This process draws on *anticipatory ethics*, like frameworks like TechEthos's Anticipatory Ethics Model [10]. Engaging diverse stakeholders (engineers, ethicists, workers, possibly the public) in scenario planning ensures a wide range of perspectives on what could go wrong and what values need protection.
2. **Human-Centric Design and Control:** An ethical framework must keep human well-being and agency at the center. This aligns with widely accepted principles such as Australia's *"Human-centered values"* in AI ethics. In practice, it means designing technology *for* and *with* the people on the front lines. Workers should be involved in the design of safety tech that will affect their jobs; their local knowledge can guide what is truly helpful versus what is overbearing. Human oversight should be maintained over automated safety systems, especially in critical decisions. For instance, rather than a black-box AI shutting down a plant on its own, a framework could require a human supervisor to confirm critical shutdown decisions (except in extreme emergency conditions). The system should be *augmentative*: AI handles complexity and routine vigilance, while humans provide judgment and moral insight. Keeping "humans in the loop" also helps preserve human skills and trust in the system.
3. **Transparency and Explainability:** Any AI or complex automation used in safety should be as transparent as possible. This is crucial for trust; people need to understand how and why a system is making the decisions it makes. An ethical integration framework would mandate explainability for safety-critical algorithms. For example, if an AI rates certain scenarios as high-risk and triggers alarms or evacuations, the criteria for those judgments should be documented and, where feasible, explainable to users and those affected. This principle is reflected in global guidelines (the EU's trustworthy AI guidelines call for transparency, and Australia's principles call for "*contestability*," the ability to challenge an AI decision). In practical terms, companies might implement *AI transparency reports* and provide channels for workers to ask questions or contest automated decisions that affect safety (e.g., "Why was my driving flagged as unsafe by the AI?"). Transparency also aids accountability: if an accident occurs, investigators and regulators must be able to audit the algorithm's role.

4. **Accountability and Governance:** Clear lines of accountability must be drawn for when technology is integrated. Ethical frameworks often insist that responsibility remains with humans, as one formulation puts it, *"people responsible for the different phases of the AI system lifecycle should be identifiable and accountable."* Concretely, this could mean assigning a *safety ethics officer* or committee to oversee new tech deployments, much like Institutional Review Boards in research. Governance policies should define who takes action if the AI behaves unexpectedly, how incidents involving technology will be reviewed, and how lessons learned will feed back into system improvements. This principle also extends to suppliers: if a third party provides an AI system (say, a machine vision for safety monitoring), the purchasing organization should ensure the vendor adheres to ethical standards and is accountable for the product's performance. Regulators and industry standards bodies are increasingly providing external governance; for example, ISO and IEEE have been developing guidelines for ethical AI design, and NIST's AI Risk Management Framework (USA) gives a model for continuous monitoring, risk mitigation, and documentation throughout an AI system's life. An organization's internal framework should align with these emerging standards, creating a *culture of accountability* around technology.
5. **Education and Training:** Ethical integration is not just about systems, but people. A new safety paradigm requires that employees, managers, and engineers are educated about both the technology and its ethical dimensions. This involves training workers to safely work with/alongside new tech (e.g., how to interact with collaborative robots, what data is being collected by their safety apps, etc.), and training leaders in the basics of AI ethics and governance. Cultivating what we might call *"technological literacy with ethics"* empowers everyone to spot issues early. For example, if a machine learning model is flagging too many false positives, a well-trained safety manager can question its validity and bias. Equally, workers should know their rights and how to voice concerns if the tech is causing undue stress or seems unsafe. Ethical frameworks like the EU's Trustworthy AI emphasize *"human agency"*; part of that is giving humans the knowledge to exercise agency when working with AI. Education ensures that the introduction of technology doesn't become a top-down imposition; however, it is a collaborative improvement with informed participants.
6. **Continuous Monitoring and Adaptation:** Even with the best planning, unforeseen ethical and safety issues will emerge once a technology is in use. Therefore, an integration framework must include ongoing monitoring, audit, and feedback loops. This means regularly reviewing the system's performance data, incident reports, and even worker feedback to catch problems early. For AI, this might involve periodic audits for bias drift or security vulnerabilities. If a new type of incident occurs (say, an AI fails to detect a hazard it should have), there should be a process to promptly investigate and update the system or procedures. In essence, treat the ethical

integration as *never "finished"*; it is an iterative process of improvement. The NIST AI risk framework and others advocate exactly this: continuous risk management, stakeholder engagement, and adaptation as conditions change. Flexibility is key; for example, if monitoring reveals that workers in an AI-enhanced process are suffering higher stress (as in the earlier warehouse example), the company must be willing to adjust workload expectations or provide more breaks, i.e., *change the system* in response to ethical reality on the ground. In regulated sectors, maintaining compliance with updated laws (like new safety regulations for digital systems) is also part of this adaptive duty.

7. **Stakeholder Inclusion and Fair Impact:** Ethical integration frameworks should ensure inclusive decision-making and fair distribution of benefits. This principle emerges from the idea of *beneficence and justice* in ethics. Technologies should *benefit* those involved and not just shift risk from one group to another. In practical terms, before rolling out, say, an AI-driven safety analytics platform, companies might hold consultations or workshops with employees, worker unions, safety committees, and possibly community representatives (if public safety is involved). This inclusive approach can highlight concerns early and also help shape technology to truly solve the problems people face on the ground. Additionally, if efficiencies from tech are gained (for example, fewer accidents and less downtime), an ethical view would suggest reinvesting some of those gains into improving worker conditions or sharing productivity benefits, rather than simply cutting costs. The "human, societal and environmental well-being" principle of Australia's AI Ethics Framework echoes this: technology integration should benefit individuals and society at large. Applying that to safety means aiming for *win-win outcomes*, safer workplaces and a supported, valued workforce, rather than safety at the cost of other human values.

By following these framework components, organizations can develop a structured approach to ethical tech integration. For example, Safe Work Australia's guidance encourages such practices, and progressive companies are already adopting internal policies along these lines. Notably, Microsoft, Google, and other tech firms have published AI ethics guidelines which include principles like fairness, reliability and safety, privacy, and inclusiveness. These mirror the points above and show a convergence toward common ethical integration strategies. The challenge is operationalizing them in each specific context, which requires commitment from leadership down to frontline employees.

GLOBAL AND AUSTRALIAN PERSPECTIVES: TOWARD A SAFE TECH CULTURE

On a global level, there is growing consensus that *technological innovation must go hand in hand with ethical safeguards*. Bodies like UNESCO have even introduced a global *Recommendation on the Ethics of AI* (2021) [11], emphasizing principles like human rights, safety, and accountability as universal standards. This reflects a worldwide recognition that without a moral compass, technology can undermine trust and safety. The World Economic Forum and other international organizations also advocate for global tech ethics standards, warning that public trust and safety are at risk if we do not set common baselines [12].

Australia provides an interesting case study of how a nation can approach this balance. The Australian government issued voluntary AI Ethics Principles in 2019 (summarized earlier) which explicitly include *safety, privacy, fairness,* and *accountability.* Building on this, Australian regulators are taking steps to enforce these ideas in critical areas. We saw how New South Wales is moving to mandate *duty of care for AI systems* in workplaces. Additionally, Australian industries like mining and healthcare have been proactive. Mining, in particular, has combined innovation with safety regulation: Australia's early adoption of autonomous mining trucks came with strict safety oversight, worker training programs, and incident investigations to ensure these systems truly delivered safety improvements. The result has been fewer traumatic injuries in mines, while regulators continue to update codes of practice for autonomous systems to address any new hazards (such as software failures or interactions between automated and manual equipment).

Australian workplaces also illustrate the importance of culture, often termed "*safety culture*," in technological ethos. A 2025 safety report (Figures 5.1 and 5.2) noted that while overall workplace fatalities in Australia have been trending down (188 in 2024, down from 203 in 2023, continuing a long-term decline), certain areas like falls from heights remained stubbornly high due to behavioral and management factors. In 2024, there were 188 workers who died from traumatic injuries while working. The frequency rate of worker fatalities was 1.3 deaths per 100,000 workers. Both of these figures are slightly lower than the five-year averages (191 deaths per year and 1.4 deaths per 100,000 workers from 2020 to 2024). Since 2014, the fatality rate has decreased 24% (from 1.7 fatalities per 100,000 workers); however, this has remained relatively steady in recent years. In 2024, Queensland recorded the highest number of fatalities (53 workers), followed by New South Wales (48 workers).

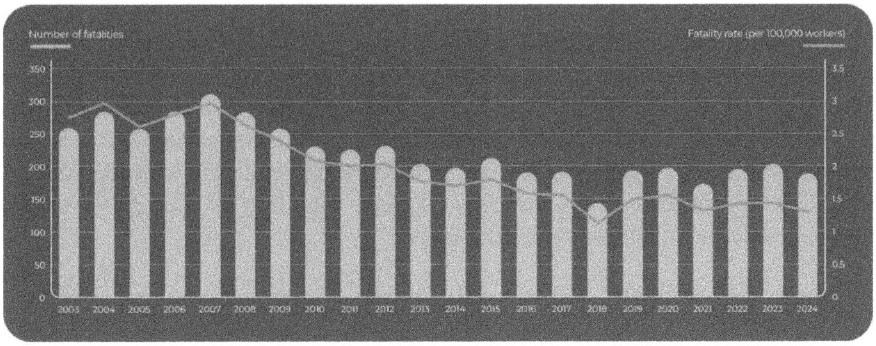

FIGURE 5.1 Australia Trends in worker fatalities, 2003 to 2024 (https://data.safeworkaustralia.gov.au).

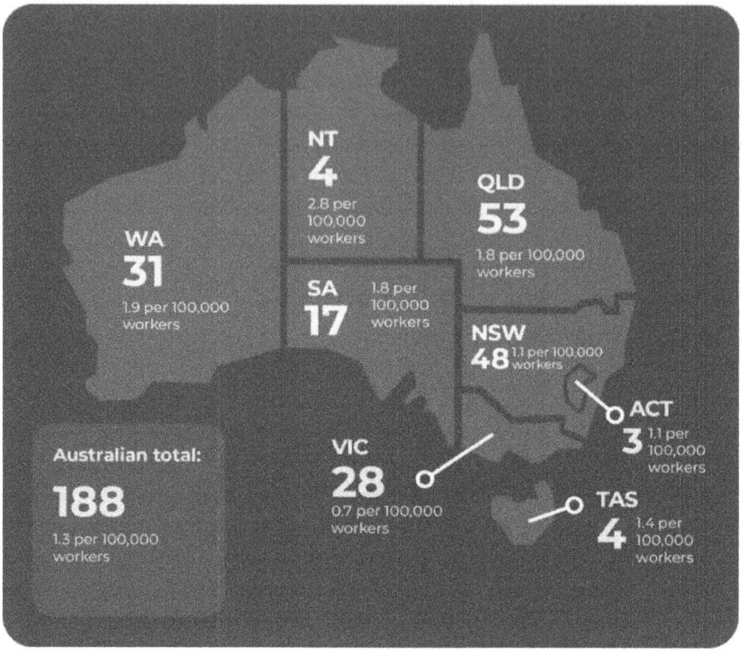

FIGURE 5.2 Australia worker fatalities by state/territory, 2024.

Northern Territory recorded the highest fatality rate (2.8 fatalities per 100,000 workers), followed by Western Australia (1.9 fatalities per 100,000 workers).

Technology alone was not a silver bullet; what's needed is *"meaningful cultures of safety"* where everyone from employers to workers actively participates in making systems (tech and non-tech) work for protection. This underscores that the ethical integration of technology is part of a broader cultural project, cultivating an ethos where safety is valued as much as efficiency.

After decades of improvement, Australia's annual workplace fatality counts have plateaued in the high-100s in recent years. The fatality rate per 100,000 workers has fallen by about 50% since 2007—a testament to better safety practices and technologies, yet new strategies are needed to break through the stagnation. Emerging technologies, ethically deployed, are seen as a key part of the next evolution in saving lives; however, equally, attention to training, enforcement, and safety culture remains critical (e.g., addressing the rise in falls from height despite available safety gear). This mix of data shows that technological gains can be real; however, they must be coupled with human-focused efforts.

CONCLUSION

"Technological ethos and safety innovations" are not opposing forces; however, complementary ones, if guided by ethics. The philosophical reflections here lead to a clear conclusion: we must foster an ethos of technology that inherently values safety, responsibility, and human dignity. As AI and automation become ever more embedded in our workplaces and daily lives, they will continue to challenge traditional safety paradigms. Our response cannot be to reject innovation, nor to embrace it blindly. Instead, we should integrate innovation ethically, meaning we design, deploy, and govern emerging technologies in line with our deepest safety values.

Practically, this means establishing strong ethical frameworks (as outlined) at organizational and societal levels. It means global cooperation on standards so that *"safe tech"* is a shared goal, much like cybersecurity is addressed today. It also means learning and adapting: ethical integration is an ongoing process of monitoring and improvement, not a one-time checklist. When done right, the rewards are great: technology can take humans out of harm's way, prevent disasters through foresight, and even augment our ethical decision-making with new tools (for example, AI could help identify unseen safety risks or ethical inconsistencies).

Ultimately, the technological ethos of the future should be one where innovation and safety are mutually reinforcing. We can imagine a future in which workplace injuries and accidents become exceedingly rare, not because risk has vanished; however, because intelligent systems and ethical practices together create an environment where *every foreseeable risk is mitigated*, and when the unforeseen strikes, resilient systems (human and machine) respond effectively. Achieving this requires a commitment to ethical integration now. As the saying goes, *"with great power comes great responsibility."* AI and automation give us great power to improve safety; it is our responsibility to wield that power with wisdom and care. By doing so, we honor both the promise of technology and the primacy of human well-being, ensuring that safety innovations truly make the world a better and safer place for all.

REFERENCES

1. Baker, T. and Welter, F., 2024. Silicon Valley entrepreneurship–Revisiting a popular dream. *Journal of Business Venturing Insights*, 21, p.e00466.
2. De Graaf, G. and Paanakker, H., 2015. Good governance: Performance values and procedural values in conflict. *The American Review of Public Administration*, 45(6), pp.635–652.
3. Hollnagel, E., 2014. *Safety-I and Safety-II: The Past and Future of Safety Management.* (1st ed.). CRC Press. https://doi.org/10.1201/9781315607511
4. Sloksnath, P., 2025. Delegating moral decisions to AI systems. Master's thesis, University of Zurich.
5. Lodge, M., 2004. Accountability and transparency in regulation: Critiques, doctrines and instruments. In *The Politics of Regulation*, pp.124–144.
6. Elmaghraby, A.S. and Losavio, M.M., 2014. Cyber security challenges in Smart Cities: Safety, security and privacy. *Journal of Advanced Research*, 5(4), pp.491–497.
7. Modi, T.B., 2023. Artificial Intelligence ethics and fairness: A study to address bias and fairness issues in AI systems, and the ethical implications of AI applications. *Revista Review Index Journal of Multidisciplinary*, 3(2), pp.24–35.
8. Fridland, E., 2014. They've lost control: Reflections on skill. *Synthese*, 191(12), pp.2729–2750.
9. Gasser, U., 2023. An EU landmark for AI governance. *Science*, 380(6651), pp.1203–1203.

10. Bhalla, N., Brooks, L., Cannizzaro, S. and Richardson, K., 2025. An ethical framework for emerging technologies: The TEAeM approach. *Journal of Ethics and Emerging Technologies*, 35(1), pp.1–16.
11. Resseguier, A. and Rodrigues, R., 2021. Ethics as attention to context: Recommendations for the ethics of artificial intelligence. *Open Research Europe*, 1, p.27.
12. Bayne, N., 1998. International economic organizations. In *Autonomous Policy Making by International Organizations*, p.195.

6 Global Narratives and Local Realities

Philosophical Approaches to Crafting Universal Safety Strategies with Cultural Sensitivity

INTRODUCTION

Globalization and rapid technological change have created transnational workforces and supply chains, bringing diverse cultures into shared safety-critical contexts. At the same time, major accidents (from industrial disasters to pandemics) have highlighted the need for robust safety strategies that transcend borders. A global narrative of safety, often framed in terms of universal human rights to health and security, provides a common language and ethical grounding. For instance, business ethicists note that respect for fundamental values (like life and fairness) extends "*beyond cultural differences*" and is rooted in a shared human nature. Many international standards (e.g., ISO 45001 on occupational health and safety) and UN guidelines espouse universal principles, reflecting the belief that certain safety obligations apply everywhere.

However, local realities can differ sharply. Each society interprets risk and safety through its cultural lens, shaped by history, religion, socio-economic conditions, and social norms. For example, some cultures emphasize collective well-being and group decision-making, while others prize individual autonomy. Power structures vary: in high power-distance societies, subordinates may defer to authorities and expect strict rules; in egalitarian cultures, workers may demand participatory safety programs. Such differences mean a one-size-fits-all safety policy can miss the mark: measures designed with one context in mind may falter or even backfire in another.

This chapter explores philosophical perspectives that address the tension between universal safety standards and culturally specific realities. Drawing on ethical theory and organizational safety literature, it demonstrates how global principles can be operationalized through culturally informed adaptation. By positioning safety as both a universal moral commitment and a context-sensitive practice, organizations can design strategies that are broadly robust while remaining locally meaningful. The chapter begins by examining core ethical foundations—contrasting universalism and relativism, as well as duty-based and consequentialist traditions—alongside

DOI: 10.1201/9781003634720-6

established safety culture theory [1]. It then analyzes key cultural dimensions and their influence on preferred safety practices and governance models. Comparative case illustrations, including regulatory contrasts between the United States and the United Kingdom, highlight practical implications for multinational contexts. Relevant cross-cultural and safety science research is synthesized throughout, with key insights consolidated in tables. The overall aim is to provide a structured framework for aligning global safety imperatives with respect for cultural diversity.

PHILOSOPHICAL FOUNDATIONS OF UNIVERSAL SAFETY

At the philosophical level, a key question is how universal should safety norms be, versus how much they should adapt to local ethics? Moral philosophy offers several lenses:

- **Moral Universalism** asserts there are basic ethical principles valid across all cultures. Under this view, certain safety-related values (e.g., the duty to prevent harm, respect for human life) are considered fundamental. Universalist theories like Kantian deontology (duty-based ethics) or basic human rights perspectives imply an absolute obligation to protect life and well-being, regardless of cultural context. For example, one might argue there is a global duty to provide safe work conditions as an application of the principle of treating people as ends in themselves. In this sense, universal ethics motivate the creation of global safety standards (e.g., *"Ensure no worker dies or is injured at work"*).
- **Consequentialism (Utilitarianism)** similarly underpins some global safety strategies [2]. It holds that actions should maximize overall welfare. A utilitarian might support measures that yield the greatest safety benefit (the greatest good for the greatest number), even if they require imposing certain practices everywhere. Public health and safety regulations (e.g., vaccine mandates in a pandemic, or universal building codes) often have a utilitarian rationale: they aim to minimize harm and promote collective well-being.
- **Communitarian and Relativist Perspectives** emphasize that moral judgments (and thus safety norms) cannot be entirely separated from cultural context [3]. Cultural relativism argues that each society's moral codes are valid for that society. In safety terms, this suggests that what is considered "*safe behavior*" or an acceptable risk may vary by culture, and imposing foreign practices without adaptation can be ineffective or disrespectful. This view encourages listening to local values: for instance, in some communities it may be more culturally resonant to frame safety as caring for the group or respecting elders, rather than as an abstract rule.
- **Virtue Ethics and Ethics of Care** focus on character, relationships, and empathy [4]. A virtue ethicist would emphasize cultivating trust, responsibility, and compassion among workers and leaders as the basis of safety. In practice, this might translate to leadership training that builds trust across cultures, rather than solely enforcing rules. Ethics of care highlights

attending to interpersonal dynamics and local relational norms (e.g., how people like to receive feedback or guidance), shaping how safety training is delivered.

Importantly, scholars point out that recognizing cultural diversity *does not necessitate full moral relativism*. As Domènec Melé explains, even amid diversity "there are basic shared values which exist beyond cultural differences". In other words, while cultures differ, the quest for justice, safety, and dignity is often shared. This suggests a middle path and upholds universal ethical commitments (to protect life and well-being) while being open to different *expressions* of those commitments in local contexts. For example, *global fraternity and solidarity* can coexist with local customs and languages [5]. Table 6.1 outlines some major philosophical approaches and how they might inform safety strategy design.

TABLE 6.1
Major Philosophical Approaches and Their Implications for Safety Strategy Design

Philosophical Approach	Key Idea	Implications for Safety Strategy
Deontological (Duty-based)	Emphasizes moral duties and rules (e.g., *"Thou shalt not kill"*).	Frames safety as a fundamental right/duty. Leads to firm, non-negotiable safety regulations (e.g., strict compliance rules, rights to a safe workplace). Emphasis on accountability regardless of consequences.
Utilitarian (Consequentialist)	Focuses on outcomes: maximize overall welfare.	Supports data-driven, outcome-oriented standards (e.g., policies that statistically reduce accidents). Encourages cost–benefit analysis of safety interventions (e.g., highest reduction in harm per resource spent). May favor uniform solutions if overall benefits are clear.
Virtue Ethics	Focuses on character, community values, and trusting relationships.	Prioritizes cultivating trust, responsibility, and a *safety culture*. Leaders model safe behavior; team discussions emphasize ethical concerns. Adapt training to local norms (e.g., role models in community).
Communitarianism	Emphasizes community values and collective well-being.	Leverages community cohesion: group-based safety initiatives, peer monitoring. Respects local decision-making processes (e.g., consensus in a village versus individual incentives).
Relativism/Pluralism	Acknowledges multiple valid moral perspectives.	Adapts safety messages to local beliefs. For example, uses local stories or symbols to teach safety. Allows flexibility (with core goals preserved). Ensures locals have a voice in defining what feels *safe*.

These approaches are not mutually exclusive; effective strategy often blends them. For instance, one can employ utilitarian analysis to choose interventions, while insisting on deontological rights to worker safety and applying them in culturally sensitive ways. The key is *reflexivity*: being aware of the philosophical stance guiding a strategy and adjusting it when engaging with different cultures.

THE GLOBAL SAFETY CULTURE NARRATIVE

Organizational safety culture theory provides a bridge between abstract philosophy and practical strategy. Since the late twentieth century, researchers have defined safety culture as the shared values, beliefs, and practices related to safety within an organization [6]. Scholars agree that a mature safety culture has three interdependent components: normative (management's policies and rules), pragmatic (observed behaviors and procedures), and anthropological (underlying beliefs and attitudes). These components are not culturally neutral. In fact, advances in safety research have integrated the notion that organizational safety culture itself is influenced by the broader national or societal culture.

Recent literature emphasizes a more unified view: rather than isolated features, safety culture is now seen as an *overarching, multifaceted concept* blending normative, pragmatic, and anthropological aspects. In this view, there is a *universal narrative* about safety: for example, the idea that management commitment, worker involvement, and continuous learning are key to safety success. However, how these ideals are interpreted depends on local context. As Edwards et al. [7] note, the intentions behind safety policies must pass through the "filter" of local values and assumptions. In other words, a top-down safety rule will have different practical effects depending on workers' shared beliefs. If local norms prioritize speed or obedience to authority, a given policy might be enforced very differently than in a context where people value open discussion and individual initiative.

An illustrative example of the universal–local dynamic is found in the *Hofstede* and *GLOBE* research on cultural dimensions. These studies identified national-level values (e.g., individualism vs collectivism, power distance, uncertainty avoidance) that correlate with workplace behaviors [8]. GLOBE researchers explicitly linked cultural dimensions to safety management strategies. For instance, in high-uncertainty-avoidance cultures, where people prefer structure to unpredictability, safety strategies tend to emphasize rigid procedures and training [9]. In contrast, low-uncertainty cultures may rely more on flexibility and employee judgment. Similarly, in low power-distance societies (where hierarchies are flat), organizations often implement decentralized decision-making and open communication (involving workers in safety decisions) [10]. High power-distance cultures might instead use formal supervision and clearly stratified roles. Table 6.2 illustrates several cultural dimensions and corresponding safety approaches.

More broadly, safety culture research confirms that the "overarching" narrative has limits. For example, a recent cross-cultural study found that while core ideals (like involvement and compliance) are widely desired, *no exact set of behaviors*

TABLE 6.2
Cultural Dimensions and Their Corresponding Safety Approaches

Cultural Dimension	Characteristic	Safety Strategy Adaptation
Uncertainty Avoidance	Preferring clear rules to manage risk.	Emphasize detailed safe work procedures and frequent training to reduce ambiguity. Clear checklists and instructions are valued.
Power Distance	Acceptance of hierarchical authority.	In high power–distance settings, supervisors enforce safety through clear commands; communication may be top-down. In low power-distance settings, encourage worker participation and "challenge authority" safely.
Collectivism (In-Group)	Strong group loyalty and teamwork.	Leverage group cohesion: team-based safety programs, peer support, and rewards for collective safety achievements. Emphasize group norms and mutual care.
Future Orientation	Planning and investing for the future.	Focus on continuous improvement: ongoing training, future-oriented risk assessments, and long-term safety goals. Encourage delayed gratification (e.g., investing in safety today).
Humane Orientation	Valuing altruism and care for others.	Encourage open communication and caring leadership. Adopt programs that prioritize employee well-being and support (e.g., stress management), reflecting a *care ethics* approach.

applies in all contexts. These emphasize that global safety frameworks must adapt: "Elements of the ideal pragmatic component of safety culture are conceivably consistent across broader cultural contexts; however, this is not meant to imply that an exact set of compliance and participation behaviors generalize to every [culture]". In practice, this means multinational firms often need to tailor their safety policies. One company might use stories and emotional appeals to engage workers (common in the United States), while another might focus on metrics and formal feedback (as seen in parts of Europe).

The global narrative of safety culture posits that leadership, communication, training, and learning are universally beneficial to safety outcomes. This is reflected in Table 6.1's emphasis on universal rights and duties, which integrates normative and cultural factors into safety outcomes. However, local cultural values moderate how these elements are interpreted and applied. Recognizing this, the rest of the chapter turns to practical strategies for bridging the global and local: designing safety programs that uphold universal ethical commitments while honoring cultural differences.

GLOBAL SAFETY STRATEGIES AND LOCAL ADAPTATION

With philosophical and cultural principles in mind, how can organizations craft safety strategies that work globally yet respect local realities? Key guidelines emerge from the literature:

- Start with Universal Goals, Then Localize: Define safety outcomes in universal terms (e.g., zero harm, fair treatment); however, involve local stakeholders in deciding *how* to achieve them. This follows the ethics of procedural justice: the ends (safety) are universal, the means should reflect local input. For example, a multinational mining company might mandate a global rule "No entry without helmet and authorization"; however, leave it to each site manager and workers' committee to design how helmets are issued and how compliance is checked day-to-day. The universal duty (helmets are required) stands; however, the policy's enactment is tailored.
- Engage Cultural Intelligence: Recent industry thought pieces emphasize Cultural Intelligence (CQ) in risk management. CQ means understanding and adapting to local norms. For safety, this involves recognizing, for instance, that in some cultures openly criticizing safety lapses is seen as rude, so feedback channels might need anonymization. In others, a public ceremony praising accident-free records might be highly motivating. Training programs should incorporate local language, examples, and even training leaders from within the culture. The Aevitium analysis on risk notes that "cultural sensitivity fosters an understanding that people from different cultural contexts might approach risk differently." In practice, companies can develop country-specific modules or use local experts to co-deliver training.
- Combine Global Standards with Local Expertise: Many organizations operate with a global health and safety management system (HSE-MS). This might include universal policies (e.g., incident investigation procedures) that derive from best practices. But to make them resonant, site managers should involve local employees in adapting the language and processes. For example, Swiss engineering standards might be introduced in a factory, a factory might adjust documentation to reflect local measurement units or clearly incorporate pictures for workers with diverse literacy levels. Studies of safety questionnaire adaptations show that even well-validated tools (like the Hospital Survey on Patient Safety Culture) need careful cultural adaptation and translation [11]. In safety strategy terms, this means piloting new initiatives in each locale and soliciting feedback before full rollout.
- Respect Local Values and Storytelling Traditions: In some contexts, narrative and communal identity carry more weight than dry metrics. US companies, for instance, often use personal stories of accident survivors to build emotional commitment to safety. In other cultures, safety messages might be woven into traditional practices or proverbs. For example, a mining company in Ghana incorporated local proverbs about strength in unity

to encourage buddy checks before work. This aligns with the ethics of care: acknowledging that how messages are framed (honoring local symbols) affects acceptance.

- Adapt Leadership and Communication Styles: The US–UK example illustrates how underlying culture shapes leadership style. In the US, the litigious environment and legalistic framework (OSHA) mean leaders emphasize compliance and often use emotional appeals. In contrast, British companies (under a principle-based law) focus on developing a safety ethos and preventive culture. Thus, a US-trained safety manager in Japan (a high power-distance, high uncertainty-avoidance culture) would need to adapt: rather than citing legal penalties, they might present safety in terms of team loyalty and national pride. Surveys suggest that truly effective global firms *"recognize these cultural and regulatory differences"* and use them to tailor strategies.

Table 6.3 compares US and UK approaches as an example of global vs local regulatory culture.

- Establish Feedback Loops: Any safety strategy, no matter how well-intended, requires evaluation. Connected sources emphasize that local workers should be able to measure and report on safety in ways that matter to them. For instance, if international benchmarks use "lost-time incidents" but local workers care more about minor injuries or near-misses, the feedback metrics should include those. Local safety committees or "cultural liaisons" (trusted employees who bridge cultures) can gather insights on whether policies are working. This is an application of Kantian respect for persons: treating workers as agents whose voices are heard. In Australia

TABLE 6.3
Comparative Overview of United States and United Kingdom Regulatory Approaches: Global Standards and Local Practices

Country	Regulatory Approach	Cultural Focus	Example in Practice
United States	Detailed, prescriptive standards (OSHA).	Compliance-driven; high litigation risk. Uses emotional storytelling to change behavior.	Emphasizes citations/ enforcement. Safety communications often share personal accident stories for impact.
United Kingdom	Broad, principle-based (Health & Safety at Work Act).	Leadership and prevention orientation [5]. Emphasizes cultural maturity and metrics.	Focuses on risk management training and data-driven monitoring (e.g., trends, safety climate surveys).

and New Zealand, the concept of *cultural safety* further reinforces this: it insists that service providers (like employers) be *accountable* to the definitions of safety given by community members. In practice, this might mean holding forums with local employees or even community elders to discuss and refine safety rules.

- Build Cultural Competence, Then Safety: Research in healthcare suggests shifting from mere "cultural competency" (knowing about other cultures) to deeper *"cultural* safety," which highlights power dynamics [12]. Translated to occupational safety, this means acknowledging that historical inequalities (e.g., colonial legacies, socioeconomic status) affect trust in management. Philosophically, it aligns with power-awareness ethics: those designing safety policies must reflect on their own cultural biases and aim for genuine partnership. Simple steps can include language translation; however, truly culturally safe programs go further, training foreign staff on respectful communication and examining whether procedures inadvertently privilege one group over another. Ultimately, cultural safety places the burden on the institution to make environments free of cultural threat, an approach that can unify global intentions with local respect.
- Leverage Universal Storytelling: One practical technique is to use stories or symbols that resonate globally *and* locally. For example, the shared value of protecting children can be a universal rallying point; each country can tie it to local festivals or media. Philosophers like Aristotle and Confucius , as noted in cross-cultural ethics discussions, found common ground in virtues like benevolence. Safety trainers might similarly find overlapping themes (e.g., "honor those who came before us by keeping each other safe") that appeal across cultures.

CASE STUDIES AND RESEARCH INSIGHTS

To ground these ideas, it is briefly considered illustrative cases and studies:

- Multinational Construction: Researchers found that a "deficit safety culture" was exacerbated by regional cultural factors [13]. Factors like inadequate training and disregard for rules (linked to economic constraints and cultural norms) contributed to high accident rates. The study concluded that improving safety in such emerging contexts requires addressing local culture and values: safety programs must account for local attitudes toward risk and authority. This supports the notion that *culture significantly influences safety behaviors* [14]. An example remedy was to involve local supervisors in crafting site-specific safety guidelines, rather than imposing external standards.
- Global Logistics Firm: A global logistics company operating encountered resistance when rolling out a centralized safety app [15]; workers (culturally

lower power–distance) were initially skeptical of mobile monitoring, preferring to report hazards in person. The company revised its approach by allowing teams to customize alert frequencies and involve worker councils in the app design, merging its universal tech solution with local process expectations. In other places (higher power–distance, more collectivist), the same app was accepted; the training sessions were adapted to include communal team exercises, tapping into the cultural emphasis on family-like solidarity. This practical example (hypothetical reflective of common practice) shows the principle: one technological tool (global narrative: digital safety) can be implemented in culturally distinct ways.
- Healthcare and Indigenous Contexts: The health sector provides a well-documented example of universal versus local safety. In New Zealand and Australia, indigenous communities have advocated for "cultural safety" in hospitals, which goes beyond checklists to address historical mistrust [16]. While medical best practices are universal (e.g., hand hygiene, checklists), the *delivery* of care is tailored: staff learn about indigenous beliefs, language, and community structures. A similar approach can apply to workplace safety among First Nations communities. For example, mining companies in Australia now engage Aboriginal elders when planning mine safety talks, framing them in ways respectful of local spiritual connections to land (thus honoring a universal safety goal via culturally sensitive means). These initiatives resonate with the chapter's theme: universal commitment (no harm to any worker) is fulfilled through local wisdom and participation.
- Corporate Training Programs: Many firms have moved toward *cultural competence* training for safety. While evidence on effectiveness is mixed, a key insight is that the highest impact comes when training is ongoing, interactive, and tied to the local context, not just one-off lectures. That aligns with the "virtue ethics" approach: developing habits and values over time. In one survey of multinational workplaces, sites that conducted regular, locally-led safety discussions (rather than solely global webinars) reported better safety climate scores. This mirrors psychological research on safety climate: the perception of an organization's commitment to safety improves when communication channels are open and culturally appropriate.

Across these cases, one central research-backed idea emerges: "For global businesses, recognizing cultural and regulatory differences is key to building cohesive, effective safety strategies that transcend borders." In other words, the *meta-strategy* is cultural awareness and adaptation. This means embedding reflexivity and learning in any global safety program.

IMPLEMENTING CULTURALLY SENSITIVE SAFETY STRATEGIES

Based on the above, practical guidelines are offered for integrating global safety narratives with local realities:

1. Assess Cultural Context. Before rolling out a new safety policy, research the local culture. Identify relevant values (e.g., individualism vs collectivism, uncertainty preference). Use tools like cultural surveys or partner with local social scientists.
2. Set Universal Principles. Allow local translation, define clear non-negotiable safety goals (e.g., *"report all hazards"*), allow local teams to decide how to phrase and enforce them. For example, one site might use union meetings to communicate rules; another might use digital alerts.
3. Engage Local Leaders and Employees. Involve respected community figures or long-tenured employees in decision-making. Workers are more likely to embrace safety initiatives if their peers or cultural role models endorse them.
4. Customize Training and Communication. Use local language, imagery, and examples in training materials. Incorporate local idioms or proverbs that relate to safety or teamwork. For example, a safety poster might feature a locally famous landmark or proverb about working together.
5. Blend Approaches. Combine global best practices (like risk assessment frameworks) with local innovations. For instance, alongside an international incident-report form, allow narrative storytelling sessions where workers from different cultures share experiences.
6. Monitor and Feedback. Establish mechanisms to gather local feedback on safety programs. This could be anonymous surveys, suggestion boxes, or regular focus groups. Adjust the strategy based on what is working or failing in each locale.
7. Build Cultural Competence. Train global safety managers not only in technical standards but also in cultural sensitivity. As one expert puts it, bridging global risk management requires high cultural intelligence to "ensure that risks are not overlooked." Encourage managers to reflect on their own biases and to seek advice from local team members.
8. Sustain Engagement. Cultures evolve, and so do organizational dynamics. Periodically revisit the cultural alignment of safety strategies. What worked five years ago may need revision today.

A concise summary of these steps is provided below:

- *Understand*: Research local cultural values and norms.
- *Define*: Establish universal safety outcomes and core policies.
- *Adapt*: Let local teams customize implementation (language, procedures, training style).
- *Engage*: Involve local staff in planning and feedback (cultural safety approach).
- *Communicate*: Use culturally resonant messages (stories, symbols, community meetings).
- *Evaluate*: Collect local data and stories to assess effectiveness; adjust as needed.

- *Train*: Develop global teams' cultural intelligence and humility in safety leadership.

Each of these guidelines is supported by research and practice. For example, the GLOBE-based analysis cited effectively embodies step 1: it highlights that understanding cultural dimensions should precede strategy design. The Aevitium risk article emphasizes the importance of steps 4–6 (training, cultural competence, feedback). Case studies of multinational firms (steps 3 and 7) show that engaging local culture leads to better safety outcomes and trust.

CONCLUSION

Crafting safety strategies that work "everywhere" without disregarding "anywhere" is a profound challenge, one that sits at the intersection of ethics, psychology, and management. This chapter has argued that universal safety strategies and cultural sensitivity are not in conflict; they are complementary if approached thoughtfully. Philosophically, one can maintain that human dignity and life are universally valuable, while pragmatically accepting that achieving a safe culture requires listening to diverse voices. The research reinforces this: broad safety ideals must be filtered through national and organizational cultures.

By integrating a pluralistic ethical outlook with empirical insights on culture, organizations can formulate *glocalized* safety programs, those guided by a global narrative of care and responsibility, yet executed in ways resonant with local realities. As recent literature on safety culture notes, organizations ultimately *reflect* the values of their societies. Thus, forging safety strategies is also an exercise in diplomacy and empathy.

In practice, this means continuously learning and adapting. It means treating global safety guidance (from international agencies, corporate headquarters, etc.) as a menu of proven options, and working with local stakeholders to pick and adapt what fits. And it means acknowledging that sometimes, the best safety solution will look different in Country A than in Country B, even if the underlying principle (say, *"everyone looks out for each other"*) is the same.

The tables and lists in this chapter provide starting points for action; real success comes from on-the-ground collaboration. As one global safety consultant observed, "there is no 'one-size-fits-all' solution for global organizations." Instead, a combination of cultural intelligence and a shared ethical commitment to protect life can guide organizations to safety strategies that are both universally principled and locally effective.

REFERENCES

1. Teck, T.S., How, L.C., Karuppiah, N. and Ho, C.J., 2018. Universalism and utilitarianism, two worlds apart or inextricably linked. *Journal of Management and Sustainability*, 8, p.13.

2. Sen, A., 1979. Utilitarianism and welfarism. *The Journal of Philosophy*, 76(9), pp.463–489.
3. Haste, H., 1996. Communitarianism and the social construction of morality. *Journal of Moral Education*, 25(1), pp.47–55.
4. Halwani, R., 2003. Care ethics and virtue ethics. *Hypatia*, 18(3), pp.161–192.
5. Sachs, A., 2012. Liberty, equality, fraternity: Bringing human solidarity back into the rights equation. *Journal of Human Rights Practice*, 4(3), pp.365–383.
6. Guldenmund, F.W., 2000. The nature of safety culture: A review of theory and research. *Safety Science*, 34(1–3), pp.215–257.
7. Edwards, L. and Veale, M., 2017. Slave to the algorithm? Why a'right to an explanation'is probably not the remedy you are looking for. *Duke Law & Technology Review*, 16, p.18.
8. Taras, V., Kirkman, B.L. and Steel, P., 2010. Examining the impact of culture's consequences: A three-decade, multilevel, meta-analytic review of Hofstede's cultural value dimensions. *Journal of Applied Psychology*, 95(3), p.405.
9. Schulman, P.R., 2020. Organizational structure and safety culture: Conceptual and practical challenges. *Safety Science*, 126, p.104669.
10. Mookherjee, D., 2006. Decentralization, hierarchies, and incentives: A mechanism design perspective. *Journal of Economic Literature*, 44(2), pp.367–390.
11. Palmieri, P.A., Leyva-Moral, J.M., Camacho-Rodriguez, D.E., Granel-Gimenez, N., Ford, E.W., Mathieson, K.M. and Leafman, J.S., 2020. Hospital survey on patient safety culture (HSOPSC): A multi-method approach for target-language instrument translation, adaptation, and validation to improve the equivalence of meaning for cross-cultural research. *BMC Nursing*, 19(1), p.23.
12. Curtis, E., Jones, R., Tipene-Leach, D., Walker, C., Loring, B., Paine, S.J. and Reid, P., 2019. Why cultural safety rather than cultural competency is required to achieve health equity: A literature review and recommended definition. *International Journal for Equity in Health*, 18(1), p.174.
13. Fayyaz, K., Shahzaib, M., Aziz, A., Irfan, M., Salah Alaloul, W. and Musarat, M.A., 2025. Cultural factors impacting health and safety (H&S) practices in a developing construction economy. *Sustainability*, 17(3), p.911.
14. Yorio, P.L., Edwards, J. and Hoeneveld, D., 2019. Safety culture across cultures. *Safety Science*, 120, pp.402–410.
15. Folkers, A. and Stenmanns, J., 2019. Logistical resistance against operations of capital: Security and protest in supply chains and finance. *Geoforum*, 100, pp.199–208.
16. Hardy, B.J., Filipenko, S., Smylie, D., Ziegler, C. and Smylie, J., 2023. Systematic review of Indigenous cultural safety training interventions for healthcare professionals in Australia, Canada, New Zealand and the United States. *BMJ Open*, 13(10), p.e073320.

7 From Harm to Harmony
A Philosophical Framework for Integrated Safety

INTRODUCTION

The energy sector, encompassing both non-renewable and renewable systems, sits at the nexus of modern society's sustainability, technology, and risk. Traditional approaches to safety in energy have often focused narrowly on preventing accidents or containing specific hazards. However, emerging challenges, from the rapid transition to low-carbon energy to the complexity of interconnected global supply networks, demand a more holistic perspective. Rather than treating safety as a series of isolated technical fixes or regulatory compliance tasks, an integrated approach seeks to embed safety into the entire energy system.

Energy provides the backbone of contemporary life; however, its production and consumption also carry profound consequences. Large-scale energy accidents, such as industrial explosions, major oil spills, or nuclear meltdowns, have shown how a single failure can devastate communities and ecosystems for generations. At the same time, the collective impact of energy use on climate change raises safety concerns of unprecedented global scope. In addition, rapid innovation, such as smart grid technologies and advanced energy storage, is creating new hazards that traditional safety models may not anticipate. These realities make clear that energy safety cannot be siloed; it must serve broader human and ecological goals.

Safety in energy cannot be understood merely as the absence of catastrophic failure, though preventing such failures remains critically important. Instead, safety must be woven into the very fabric of energy systems at every scale, from local communities to international grids. An integrated safety paradigm recognizes the interdependencies among technology, environment, and society. It acknowledges that energy operations have broad ramifications: for public health, environmental sustainability, social equity, and even geopolitical stability. From this viewpoint, achieving safety is akin to achieving a kind of harmony between energy provision and the values of society.

The chapters preceding this one have examined technical risk analyses, policy frameworks, and ethical case studies across various industries and sectors. Here we draw those threads together in the specific context of energy. This concluding chapter is a philosophical synthesis intended to weave concepts into a coherent whole. The energy sector's challenges are not only technical; they are deeply social and ethical. Accordingly, we combine systems thinking, safety ethics, and epistemological

DOI: 10.1201/9781003634720-7

integration to propose new perspectives and methods for understanding and managing safety in energy.

The journey from harm to harmony is both practical and philosophical. It requires seeing safety not as a constraint on progress; however, as an opportunity to innovate and align energy systems with broader human and ecological well-being. This chapter offers a conceptual compass for that journey: to understand safety in a holistic way, to bridge theory and practice, and to chart a path toward energy systems that are as harmonious as they are resilient and sustainable.

SAFETY AS A SYSTEMIC PROPERTY

In the energy sector, safety must be understood as an emergent property of the entire socio-technical system rather than a simple technical parameter. Complex interactions among equipment, human operators, organizations, and natural environments shape outcomes in unexpected ways. For example, analysis of past energy disasters indicates that technical failures often combine with organizational or environmental factors to create widespread damage. In this view, no single device or rule guarantees safety; instead, safety emerges when the whole system of technology, people, and institutions functions harmoniously.

Key characteristics of an integrated safety paradigm include: Interconnectedness: Every component and process influences others, so small changes or failures can cascade. Emergence: System-level behaviors, such as resilience or instability, cannot be predicted by examining parts in isolation. Context-dependence: The same technology or process may be safe in one setting; however, unsafe in another, depending on environmental or social conditions. Dynamism: Energy systems constantly evolve (with new technologies, market shifts, and policy changes), so safety must adapt continuously. Normativity: Safety is linked to values; what counts as safe or desirable involves ethical and cultural judgments, not just technical specifications.

Figure 7.1 illustrates Integrated Safety as an emergent property of an interconnected energy system. At the center, safety is not treated as an isolated technical outcome but as the result of continuous interactions among four key domains: Technology, Environment, Society, and Values and Norms. Each outer component influences the others through feedback loops. Technological systems affect environmental conditions; environmental impacts shape social responses; societal expectations influence norms and regulations; and values, in turn, guide technological development. The circular arrows emphasize that these relationships are dynamic and ongoing.

To clarify this shift in perspective, Table 7.1 contrasts traditional safety thinking with an integrated, harmony-oriented approach tailored to energy systems.

In philosophical terms, viewing safety as systemic aligns it with concepts of resilience and sustainability. A truly safe energy system is one that can absorb shocks and reorganize without collapsing. This resilience-focused perspective emphasizes not just preventing accidents; however, also designing networks that can adapt to and recover from disturbances. A system that is safe today may not remain so if it cannot evolve; continuous adaptation and alignment across all levels are essential.

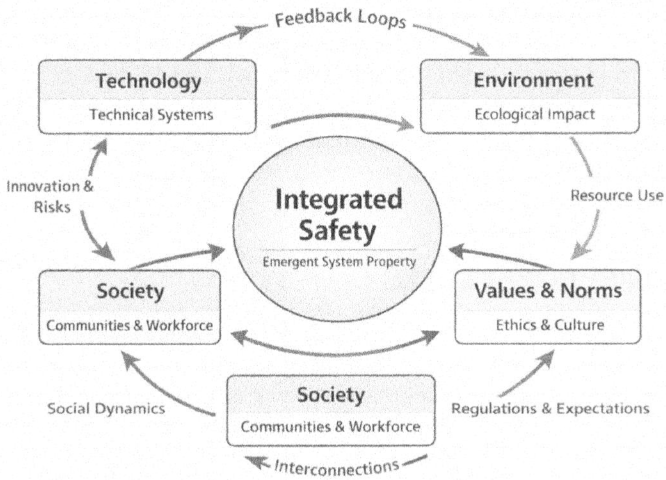

FIGURE 7.1 A conceptual diagram illustrating safety as an emergent property of an interconnected energy system. It might show feedback loops between technology, environment, society, and norms.

TABLE 7.1
Traditional vs Integrated Safety Approaches in Energy

Aspect	Traditional Safety (Harm-Avoidance)	Integrated Safety (Harmony-Oriented)
Temporal Focus	Reactive response after incidents; emphasis on investigation and corrective action	Proactive anticipation of risks; resilience-building embedded from planning and design stages
System View	Reductionist focus on isolated components, hazards, or failure points	Holistic perspective considering the entire socio-technical-ecological energy system
Objective	Prevent accidents and ensure regulatory compliance	Align safety with environmental sustainability, social responsibility, and operational performance
Knowledge Basis	Reliance on technical risk models, quantitative data, and past incidents	Integration of scientific analysis, local knowledge, stakeholder perspectives, and ethical reflection
Governance	Fragmented oversight across agencies or organizational silos	Coordinated and collaborative governance across sectors, disciplines, and communities

Systems thinking allows us to draw further analogies. Just as a healthy forest depends on the balance of all its trees and species, an energy system's safety depends on the condition of every part. This means fostering organizational learning and knowledge-sharing (for instance, through industry-wide incident reviews) as much as building robust equipment. It also means investing in redundancy, backup generators, spare lines, and emergency reserves that can isolate and mitigate failures before they escalate. In short, safety as a system property is about nurturing a harmonized, living system rather than applying isolated fixes.

SYSTEMS THINKING IN ENERGY SAFETY

Systems thinking reminds us that energy systems are deeply interconnected networks of technology, people, and nature. An energy plant is not just machinery; it is part of local economies, regulatory environments, and ecological cycles. Recognizing this, a systems approach emphasizes how change propagates through feedback loops and how unexpected interactions can give rise to new risks or opportunities. For example, increasing renewable energy capacity may introduce variability into the grid, creating new feedback loops that must be managed carefully. In another example, the North American blackout of 2003 began with a few local faults that cascaded through the grid, showing how minor events can have continent-wide impacts. Systems thinking would aim to anticipate such chain reactions by mapping and modeling potential feedbacks in advance.

Key elements of systems thinking applied to energy include:

Holism: Viewing the energy system as a whole, rather than focusing on individual parts. Safety emerges from the interactions of generation, distribution, consumption, and all supporting institutions.

Feedback Loops: Acknowledge that actions (like new regulations or technology adoption) can loop back to affect behavior and system stability. For instance, a safety incident can change public perception, which in turn shapes policy and future designs.

Nonlinearity: Small disruptions can lead to disproportionately large effects (e.g., a minor equipment failure causing a widespread blackout through cascading failures). This requires preparation beyond linear risk estimates.

Emergence: New safety challenges can arise unpredictably (e.g., cyber-physical interactions in a smart grid) that are not foreseeable by examining subsystems alone.

Multiple Scales: Consider safety at different levels, from component maintenance up to national energy security, as well as across time horizons from immediate operations to long-term sustainability.

Figure 7.2 presents a conceptual systems diagram of the energy sector, illustrating how safety emerges from the dynamic interactions among generation, grid infrastructure, consumers, regulations, and the environment. Rather than depicting isolated

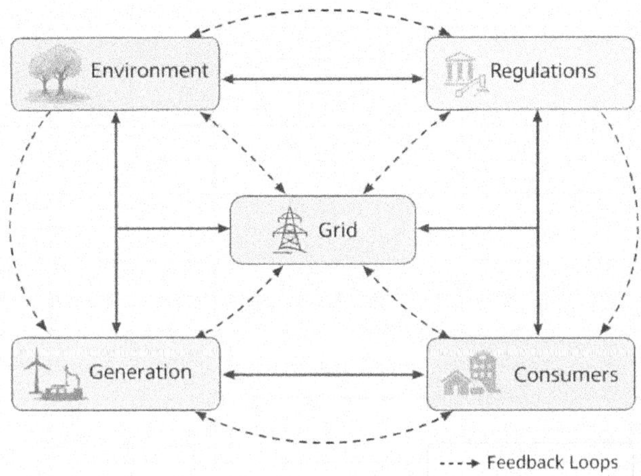

FIGURE 7.2 A conceptual systems diagram of the energy sector, showing interconnected nodes such as power plants, grids, consumers, regulations, and the environment, with feedback loops highlighted.

components, the figure emphasizes interconnected nodes and feedback loops that shape system behavior over time. The diagram highlights key systems thinking elements: holism (the energy sector as an integrated whole), feedback loops (e.g., incidents influencing public perception and regulatory change), nonlinearity (cascading effects across the grid), emergence (new risks arising from complex interactions such as cyber-physical integration), and multiple scales (from component reliability to national energy security and long-term sustainability). Together, these interactions demonstrate that safety is not a fixed attribute but an evolving property of the entire socio-technical energy system.

To illustrate, Table 7.2 outlines how systems thinking concepts translate into energy safety implications.

Systems thinking grounds safety in the real-world complexity of energy. It points to resilience engineering, adaptive management, and continuous learning as practical priorities. For example, anticipating nonlinear effects has led some grid operators to conduct "stress tests" of the network under extreme scenarios. Considering feedback loops has inspired the development of interactive control systems that adjust in real time. By emphasizing redundancy and flexibility (such as multiple supply routes or backup systems), a systems lens helps ensure that no single point of failure can jeopardize the whole. In this way, energy safety becomes not just a theoretical concern; however, a practical one: decision-makers must constantly scan the system for hidden interconnections and adapt policies accordingly.

TABLE 7.2
Systems Thinking Elements and Their Implications for Energy Safety

Systems Element	Implication for Safety
Holistic Perspective	Safety assessments should simultaneously address technical, organizational, social, and environmental dimensions of the energy system.
Feedback Loops	Monitor how safety interventions influence broader social and economic dynamics, such as public trust, investment patterns, and market stability.
Nonlinearity	Prepare for low-probability but high-consequence events by strengthening system robustness and coordinated emergency response capacity.
Emergence	Promote adaptive management and continuous learning to address unforeseen risks arising from complex system interactions.
Multiple Scales	Ensure alignment of safety strategies across different levels, from facility operations to regional grids and global energy governance.

ETHICAL FOUNDATIONS OF SAFETY IN ENERGY

Safety is inherently an ethical concept because it concerns the protection of people, communities, and the environment. In the energy sector, ethical considerations shape how we define and pursue safety. For example, engineers and policymakers must weigh the trade-off between energy benefits (power, jobs, development) and potential harms (accidents, pollution). A principled approach to safety demands that no group is unduly burdened by risks and that we honor our duty to future generations and to the living environment.

Key ethical principles relevant to energy safety include:

Precaution: Anticipate and mitigate unknown risks (e.g., regarding emerging technologies or climate impacts) rather than waiting for proof of harm.

Justice and Equity: Ensure fair distribution of risk and benefit. For example, preventing accidents in marginalized communities and enabling equitable access to safe energy.

Transparency: Make information about risks, safety measures, and incidents openly available. This builds trust and allows stakeholders to participate in safety decisions.

Responsibility: Accept duty of care for all those affected by energy activities, including downstream communities, workers, and the global ecosystem.

Intergenerational Equity: Commit to long-term safety by not transferring risks to future generations (for example, by avoiding leaving hazardous waste or exacerbating climate change).

Solidarity and Sustainability: Recognize that safety contributes to the long-term well-being of society and the planet. Invest in safe practices that support sustainable energy transitions rather than short-term gains.

Energy safety ethics demand that technology development be guided by values as much as by cost–benefit calculations. For example, an energy project might be

TABLE 7.3
Ethical Principles and Their Applications in Energy Safety

Ethical Principle	Application in Energy Safety
Precautionary Principle	Apply conservative safety margins, rigorous testing, and contingency planning when deploying emerging technologies such as carbon capture and hydrogen systems.
Justice and Equity	Prioritize risk reduction measures for vulnerable or high-exposure communities and ensure fair compensation and recovery mechanisms following incidents.
Transparency	Disclose safety performance data openly, communicate risks clearly, and incorporate public and stakeholder input into safety standards and decision-making.
Responsibility to Future Generations	Integrate long-term environmental, climate, and public health impacts into safety assessments to prevent irreversible harm or unsafe resource depletion.
Sustainability	Design and operate energy infrastructure for durability, resource efficiency, and minimal ecological disruption, linking system safety with environmental resilience.

technically robust; however, ethically flawed if it proceeds without community consent or neglects environmental stewardship. Table 7.3 illustrates how some ethical principles translate into concrete safety commitments in energy contexts.

A system may be technically sound yet ethically flawed if it imposes undue harm on certain populations or ignores broader impacts. Integrating ethical values means asking not only *"What can go wrong?"* however, also *"What ought to happen?"* This can lead to practices like stakeholder engagement and inclusion of social impacts in safety metrics. For instance, climate change highlights that emitting greenhouse gases is a safety risk for future people; taking responsibility might involve requiring carbon-neutral designs or just-transition plans for affected workers. In short, ethics expands safety from a narrow engineering problem to a question of justice and care, ensuring that energy systems serve the common good.

EPISTEMOLOGICAL INTEGRATION IN ENERGY SAFETY

Epistemology deals with the nature and scope of knowledge, and in the context of energy safety, it implies bringing together different ways of knowing. Engineers may rely on mathematical models and historical data, while community members may contribute experiential knowledge about local hazards. Scientists bring quantitative analysis, whereas indigenous custodians bring value-driven insights about the land. Integrating these epistemologies expands our understanding of safety. For example, a community living near a plant may notice subtle warning signs (smoke, noise, odors) before engineers detect a technical failure. Without integrating that knowledge, the system might miss early indicators of risk.

Key aspects of epistemological integration include:

Multi-source Information: Combining statistical risk assessments with qualitative data from case studies, oral histories, and social surveys to form a richer picture of hazards. **Uncertainty Management:** Recognizing that knowledge is never perfect. Diverse perspectives help estimate uncertainties and reduce blind spots in safety planning.

Co-production of Knowledge: Bringing stakeholders into the process of creating models or setting safety goals. This might involve participatory scenario workshops, joint development of safety indicators, or citizen science monitoring of environmental effects.

Learning Systems: Implementing mechanisms for continuous feedback and learning. For instance, anonymized near-miss reporting systems, adaptive policy reviews, and open data platforms ensure new information is integrated promptly.

Epistemic Justice: Ensuring that marginalized voices are heard in defining what is safe and acceptable. This challenges experts to consider ethical and cultural dimensions of knowledge.

Table 7.4 shows some key knowledge domains and strategies for integrating them in energy safety practice.

By weaving together these epistemologies, safety becomes not just a technical checklist; however, a living consensus. This epistemological integration aligns knowledge production with systems and ethics considerations, ensuring that both facts and values inform energy safety. In high-stakes, uncertain contexts, often called "post-normal" science, diversity of perspectives is an asset rather than a complication. In practice, this might take the form of joint risk assessment panels, citizen–science networks, or adaptive codes of practice that evolve with new insights.

TABLE 7.4
Knowledge Domains and Integration Approaches in Energy Safety

Knowledge Domain	Integration Strategy
Scientific / Technical	Apply multi-model risk analysis, openly share datasets, and integrate uncertainty analysis with forward-looking scenario planning.
Local / Traditional	Organize community workshops and field engagement; incorporate indigenous and experiential knowledge into environmental and safety assessments.
Operational / Managerial	Maintain transparent incident and near-miss reporting systems, ensure continuous training, and systematically transfer lessons learned across projects.
Regulatory / Policy	Develop multi-criteria decision-support tools and embed structured public consultation into the development of safety standards.
Ethical / Cultural	Integrate value-based criteria such as justice and sustainability into safety objectives and involve ethicists or social scientists in risk evaluation processes.

Such integration democratizes safety: engineers, citizens, and experts learn together and build a more robust understanding of risk.

BRIDGING THEORY AND PRACTICE: NEW METHODS AND APPROACHES

Moving from conceptual understanding to action requires novel methods that integrate multiple perspectives. Below we outline some emerging approaches that bridge theory and practice in energy safety:

- **Integrative Risk Assessments:** Go beyond single-hazard analysis by modeling how different risks interact (e.g., technical, environmental, and social factors) using techniques like network analysis or agent-based models. This reveals complex scenarios that linear risk tools might miss.
- **Participatory Simulation and Scenario Planning:** Engage diverse stakeholders in creating future scenarios. For example, run collaborative simulations of the grid under extreme weather, so that engineers, community members, and policymakers develop a shared understanding of vulnerabilities.
- **Safety by Design:** Inspired by ideas like "privacy by design," this approach embeds safety into all stages of development. It might include mandatory safety and ethics review boards for new energy projects or co-design sessions where safety experts and community representatives map out infrastructure plans together.
- **Resilience Dashboards:** Develop indicators of systemic health (e.g., recovery time after outages, redundancy levels) and implement real-time dashboards. Operators and regulators can then monitor not just failures; however, overall system harmony, enabling early intervention.
- **Continuous Learning Systems:** Institutionalize learning loops, such as anonymized sharing of near-miss data across organizations or "lessons learned" repositories accessible to all. This turns operational experience into a collective knowledge base that evolves safety practices.
- **Adaptive Governance:** Create regulatory sandboxes or iterative permitting processes where projects meet safety requirements and then evolve. Regulators work with developers to adjust safety measures as new data emerge, rather than using rigid one-time approvals.
- **Cross-Industry Learning:** Adapt proven safety innovations from other domains. For example, energy operators could adopt safety checklists and "just culture" practices from aviation, or collaborative incident review procedures from the chemical industry.
- **Ethical Impact Assessment:** Parallel to environmental impact statements, require anticipatory assessments of how projects might affect human and ecological well-being. This integrates ethical reflection into planning and holds developers to broader standards of care.

Table 7.5 pairs some of these approaches with their core ideas and intended outcomes.

These methods create feedback between theory and practice. For example, insights from participatory simulations might reveal new risk factors, prompting modelers to refine their analysis. Conversely, real-world data from resilience dashboards feed back into scholarly understanding of safety. In each case, tools and processes must be anchored in trust and communication. For instance, a digital simulation workshop not only produces a scenario model; it also builds relationships among participants as they deliberate solutions. In every method, technology supports; however, it does not replace the human process of designing safety.

Implementing these methods often requires new roles and institutions: risk communication officers, safety innovation labs, or community liaison positions. It also invites rethinking professional training. Engineers and policymakers might work side by side with sociologists and ethicists to co-create safety solutions, breaking down the theory-practice divide. These bridging approaches lay the groundwork for the implementation framework that follows.

TABLE 7.5
Proposed Methods and Approaches for Integrated Energy Safety

Method / Approach	Description	Intended Outcome
Integrative Risk Assessment	Apply advanced modeling techniques (e.g., network analysis, AI-based simulations) to evaluate multiple interacting hazards and cross-sector dependencies.	Holistic understanding of complex and cascading risk scenarios.
Participatory Simulation	Conduct collaborative scenario workshops using digital tools such as system dynamics models or virtual reality environments.	Development of shared mental models and informed stakeholder input.
Safety by Design	Integrate safety considerations directly into project development through co-design sessions, structured hazard reviews, and ethical evaluation.	Infrastructure and systems inherently aligned with safety objectives.
Resilience Dashboard	Implement real-time monitoring platforms that track key indicators of system performance, reliability, and safety.	Early detection of emerging issues and enhanced adaptive control.
Knowledge Co-Production	Foster continuous collaboration between experts, communities, and other stakeholders through joint monitoring and citizen science initiatives.	Inclusive policies that reflect diverse knowledge sources.
Adaptive Regulation	Establish flexible regulatory frameworks that evolve with new evidence, including experimental approaches such as regulatory sandboxes.	Governance systems capable of learning and continuous improvement.

PRACTICAL IMPLICATIONS AND IMPLEMENTATION FRAMEWORK

Translating integrated safety into reality requires coordinated action across many levels of governance and society. Below we outline key areas and actions:

- **Integrated Governance:** Establish cross-sectoral safety councils or agencies that bring together energy regulators, environmental authorities, and community representatives. Develop policies that mandate system-level safety assessments and integrate energy planning with safety criteria.
- **Safety Culture and Education:** Equip engineers, managers, and policymakers with systems thinking and ethical training. Encourage interdisciplinary teams and continuous professional development. Cultivate an organizational culture where safety and shared values are routinely addressed.
- **Standards and Tools:** Create new guidelines reflecting the integrated approach. For example, develop safety checklists that combine technical hazards with social and ecological considerations. Promote digital tools (like interactive dashboards) that make complex risk information accessible and actionable.
- **Local Capacity:** Strengthen community-based monitoring and response capabilities. This could involve funding local safety research, supporting worker involvement in safety committees, and ensuring that local knowledge is part of emergency planning and day-to-day operations.
- **Community Engagement:** Institutionalize public participation in safety governance. Form advisory boards with local communities, indigenous groups, and NGOs to review energy projects. Use transparent communication (e.g., open data portals, town halls) to build trust and accountability.
- **Metrics and Monitoring:** Define comprehensive safety metrics that include not only incident rates but also system resilience and social well-being. Track measures such as recovery time after disruptions or levels of community preparedness, and use these indicators in decision-making and reporting.
- **International Collaboration:** Share best practices globally, since many energy technologies (like nuclear power or climate-related impacts) have international implications. Support collaborative research initiatives and work toward harmonized safety standards where possible.

Implementing these pillars involves both top-down and bottom-up processes. Policymakers and companies must commit to integrated safety goals; however, equally important are grassroots initiatives that hold the system accountable. For example, citizens and NGOs can use public safety data to advocate improvements, creating a feedback loop to decision makers. Similarly, schools and universities can incorporate integrated safety principles into curricula, producing future professionals attuned to these ideas.

TABLE 7.6
Pillars of an Integrated Energy Safety Framework

Pillar	Actions / Components
Governance and Policy	Establish cross-agency safety councils, promote integrated legislation, and implement adaptive permitting frameworks.
Design and Engineering	Apply safety-by-design principles, develop modular and redundant system architectures, and embed resilience planning into infrastructure development.
Culture and Education	Provide systems-safety training, promote ethical leadership programs, and conduct regular safety culture assessments.
Community and Equity	Create stakeholder advisory boards, perform fairness and equity impact analyses, and ensure transparent and inclusive engagement processes.
Data and Metrics	Develop shared safety databases, deploy real-time monitoring systems, and use holistic performance indicators to track system health.

In practice, implementation might also include tying financial incentives (like insurance premiums or project funding) to safety performance and ethical metrics. On the ground, it means joint drills with communities, worker-led safety committees that have real influence, and public education programs about energy risks and responses. Success should be measured not only by reduced accident rates but also by metrics such as recovery times, environmental health, and public trust. Ultimately, progress depends on embedding safety into the organizational DNA of every energy enterprise. This requires reforming policy, building capacity, and fostering a culture of continuous improvement. Leaders at all levels, from industry CEOs to community organizers, must champion the vision and iterate on the framework as new challenges arise.

CONCLUSION

This chapter has presented a comprehensive philosophical framework for viewing safety in energy as a holistic, integrated system property. We have moved beyond narrow hazard control to a vision where safety is about harmony, between technology, people, and the environment. By combining systems thinking, ethics, and knowledge integration, we see that safety becomes a guiding purpose rather than just a constraint.

In practice, this framework suggests that energy projects and policies should be designed with deep anticipation of ripple effects, with ethical values guiding decisions, and with all stakeholders engaged in defining what safety means. Methods like integrative risk modeling, participatory scenario workshops, resilience dashboards, and adaptive governance can help to realize these ideas. The tables and figures throughout have sketched how such an approach might be organized; however, the real work lies in building the institutions, relationships, and cultures around these principles.

For example, imagine a coastal community installing a new wind farm. An integrated safety approach would not only ensure turbine reliability, but also anticipate storm-surge flooding (a climate-related hazard) and analyze social factors like local acceptance. Engineers, ecologists, and community members might co-develop emergency plans and monitoring systems. Residents' knowledge of local weather patterns and fishery practices would complement technical models. This co-designed safety plan exemplifies how theory and practice can unite under the harm-to-harmony ethos.

Crucially, integrated safety requires continual learning and humility. Energy systems evolve and new challenges emerge; societal values shift. Our frameworks must therefore remain flexible and adaptive. Future engineers, regulators, and citizens must carry forward the mindset that safety involves everyone's knowledge and responsibility. The framework presented here should be seen as a living synthesis that will grow with future experience and innovation.

Finally, integrated safety in energy is inherently interdisciplinary and transnational. Since climate change and energy markets know no borders, this framework emphasizes global dialogue. International agencies and NGOs can facilitate sharing of integrated safety innovations, from nuclear regulatory cooperation to community resilience networks. In this way, energy safety becomes part of humanity's common agenda, linking ethical stewardship with scientific and technological collaboration.

In conclusion, the journey from harm to harmony is ongoing. It calls on us to transcend silos, to see energy not only as power and infrastructure; however, as a shared human endeavor that must respect both scientific realities and moral imperatives. When safety is treated as integral to the very fabric of energy systems, when every innovation is pursued with care for people and planet, we move closer to a world where energy truly serves the common good. This vision is demanding; however, profound. It invites us to cultivate technologies and institutions that reflect our highest aspirations for well-being. In so doing, the energy sector can lead society into a future where security, sustainability, and harmony converge.

Index

www.ingramcontent.com/pod-product-compliance
Ingram Content Group UK Ltd.
Pitfield, Milton Keynes, MK11 3LW, UK
UKHW022122190926
475702UK00013B/1587
* 9 7 8 1 0 4 1 0 5 1 9 4 7 *